AF531561

ECONOMIC BOTANY

ECONOMIC BOTANY

By
Dr. Pooja
Department of Botany
R.C.C. College
Ghaziabad (U.P.)

DISCOVERY PUBLISHING HOUSE
NEW DELHI-110002

Published by:

DISCOVERY PUBLISHING HOUSE PVT. LTD.
4383/4B, Ansari Road, Darya Ganj
New Delhi-110 002 (India)
Phone : +91-11-23279245; 23253475; 43596065
E-mail : discoverybooksindia@gmail.com
discoverypublishinghouse@gmail.com
web : www.discoverypublishinggroup.com

***First Published:* 2005**

***Reprinted:* 2021**

ISBN: 978-81-7141-956-2

Economic Botany

Printed at:
Infinity Imaging Systems
Delhi

Preface

Animals have to depend directly or indirectly upon plants. Ever increasing population sizes have made us think about ways to feed large numbers of people, most of whom live under substandard conditions. At the same time, the natural or organic food movement has necessitated that individuals become educated consumers of plant products. As a consequence, interest in studying plants of actual or potential use in feeding, clothing, housing, and warming human kind has arisen. The present title provides a detailed coverage of the major uses of plants. To keep the text readable and relevant, efforts have been made to provide a balanced treatment of the plants discussed by including aspects of history, morphology, chemistry, and modern usage.

To make the work more comprehensive and informative, author has consulted many authoritative books, research journals, abstracts, monographs etc. Author is grateful to all those great scholars whose work are cited or substantially reproduced.

There can be no claim to originality except in the manner of treatment and much of the information has been obtained from the books and scientific journals available in the different libraries.

The author expresses her thanks to her friends and colleagues whose continue inspirations have initiated her to bring out this book.

The author is painfully aware of the shortcomings, errors and misprints that have crept in, and will be greateful to receive suggestion for improvement of the next edition from all the readers.

The author expresses her gratitute to Mr. Wasan and staff of M/s Discovery Publishing House for their whole hearted co-operation in the publication of this book.

Author

Contents

1

ORIGINS OF AGRICULTURE

Farming is a relatively recent innovation in man's world. Humans (Homo sapiens sapiens) have existed for an estimated 100,000 years, but they have practiced widespread cultivation for less than a tenth of that time. A series of questions arises when we discuss the origin of agriculture, when, where, how and why people first began to cultivate plants on a permanent basis, have intrigued man for thousands of years. While we cannot hope to do justice to all of the voluminous and scholarly work that has addressed questions about the origins of agriculture, we present in this chapter the most prominent theories of the past and present. Implict in all of the questions posed above is the assumption that humans, at some point which we can more or less define, turned from a nonagricultural to an agricultural way of life. To be precise in our use of terms, Agriculture can be defined as the caring for, or the cultivation of, plants. As Edgar Anderson and Jack Harlan have pointed out, this caring for plants can be minimal or intensive, ranging from the encouraging of essentially wild individuals to the careful planting and rearing of selectid specie.

In our discussion of the origins of agriculture, we are really talking about the adoption of agriculture as a way of life. Domestication a word often confused with cultivation is used to refer to a more complicate process than cultivation which involves the genetic alteration of plants brought about by the activities of humans. In most cases, domestication follows closely on the heels of cultivation since any selection of individuals (for example, the use of certain individuals for seed sources) would lead to genetic changes in the cultivated plants. In some cases, plants have been so altered by domestication that it is no longer possible to link them precisely with a particular wild species. Once we have discussed ideas about the origin of agriculture in general, we briefly turn our attention to suggestions about places where important crop species were first domesticated.

Time Frame

We can gegin with perhaps the easiet question: When did people first consciously begin to care for, rather than simply gather, plants? or when did man first become a farmer rather than a hunter-gatherere? This question is "easiest" only because ther are methods for dating traces of prehistoric agriculture and most workers are willing to accept the oldest date until an older one is demonstrated. What are the ways of finding and placing a date on evidences of early agriculture? Naturally, the first step in the process is to find fossil human encampments or resting places. Plant remain, tools used for cultivation and harvesting gives the evidence that agriculture was being practiced.

An analysis showing that the morphology of the fossilized plant material in digs is different from wild forms in nearby areas is considered evidence that humans altered, or domesticated, the material. Plant and animal remains for such analyses are generally found in garbage dumps, strewn around ancient hearths, or as pollen found on pots or pot fragments around the sites. Pollen grains are particulary useful for identifying remains for two reasons. First, they are exceedingly resistant to decomposition, persisting unchanged in deposits for millions of years. Second, many pollen grains have charactristic sculpturing that allows the botanist to determine the species or genus from which they came. Charred seeds or plants parts can also often be identified by comparing gross or cellular features with those of modern counterparts.

Dating is usually done by the carbon 14 method. This methods involves the natural occurance of a carbon isotope that decays at a known rate. All living organism are made up of carbon atoms. The usual form of carbon has molecular weight of 12 but other atoms reacting chemically like carbon also occur ^{14}C is one of them has a molecular weight of 14. This isotope is continuously formed in the atmosphere by the interaction of comsic rays and nitrogen. It slowly degenerates with a half-life of abour 5,700 years back into nitrogen. The proces of ^{14}C formation and disintegration has been taking place for so long that an equilibrium between ^{14}C and ^{12}C was reached millions of years ago. When an organism is alive, it is constantly incorporating carbon into its body. In the case of plants, most of the carbon incorporated, or "fixed," is taken into the plant in the form of carbon dioxide. Since ^{14}C and ^{12}C occur in a constant ration in the atmosphere, they are incorporated into plant tissues (with some selectivity) in this ratio. Once a plant dies, or is collected by humans further incorporation of ^{14}C ceases. After this time only decay of the ^{14}C into nitrogen occur. By measuring the ratios of ^{12}C and ^{14}C present in a piece of organic fossil material, a formula can be used to calculate the time since the death of the organism.

The method can be used even for cooked or burned organic material since neither heating nor charring alters the ^{12}C to ^{14}C ratios. Although ^{14}C method is very simple, accurate dating requires extra care and the use of uncontaminated material, and provides reliable dates only for for ages between 500 and 50,000 years before present (ybp). However, this time span is perfectly adequate for dates of most archaeological sites. What have data from such digs told us?

The earliest dates firmly ascribed to evidences of agriculture are between 8,000 and 10,500 years ago. Until recently, all dates over 12,000 years old appeared to be from sites that had been inhabited by hunter-gatherers. In 1797, an exciting report appeared which described a barley grain with features of domesticated kernel. This grain was found in deposits dated to be 18,500 years old, about 8,000 years older than any previous evidence of agriculture. It now appears what the barley kernel was deposited from younger sediments into a mixture of old fossilized material. Still have no strong evidence which proves firm dates for significant agriculture before 10,500 years before present, and we will assume only that by some time between 9,000 and 10,000 ybp, were humans unquestionably practicing agriculture. It is also unquestionable, on the basis of numerous archaeological sites, that by 2,000 years ago, most of the major civilizations of the world practiced agriculture.

It was during this 8,000 year period from 10,000 ybp to 8,000 ybp that man's entire way of life and the course of civilization changed. When the people became dependent on their own production of foods they rarely reversed the process to return to gathering as a means of substance.

A Cardle for Agriculture?

The earliest dated indicating the practice of agriculture come from archeological sites located aroudn the eastern edge of the Mediterranean Sea (Syria, Israel, Jordan, etc., and now, possibly Egpt. This region has been variously labeled as the Cradle of Agriculture or the Fertile Crescent. To us today, this looks like an unlikely place to begin cultivating plants. The terrain is hilly and rocky, the soil comparatively poor, and the climate arid. This area has provided the oldest firm dates for the practice of agriculture, in many other parts of the world the cultivation seems to have begun at nearly the same time.

In the New World, fossil pumpkins and gourds suggest some sort of cultivation between 7,000 and 9,000 years ago. In Asia, dates from Spirit Cave (Thailand) imply that cultivation here also began about 9,000 years ago. Cultivation probably began about the same time or only slightly later in northern Africa, but the scanty data available at the present tiem indicate only that agriculture was not adopted south of the Sahara until much later. Still, with dates from the eastern Mediterranean Sea, southeast Asia, and Central America so close in time the question is not really where did agriculture begin, but whether it began in one place and the knowledge of its practice diffused around the world or whether it was independently adopted in seveal places inthe world. Some authors believed and defined a diffusionist view, while majority of anthropologists, and archaeologists concerned with this subject believe that agriculture was independently adopted in many places around the world.

Of Myths and Man

How did humans learn to cultivate plants? Both oral and written history are full of speculations of how people first came to know about agriculture. In the Old World, ancient cultures around the Mediterranean favoured goddesses as the purveyors of the secrets of agriculture. In ancient Egypt, two of the five children of Geb and Nun (the second couple created by Ra), Isis and Osiris, married one another and the pair took over the ruling of Egypt when Geb abdicated his throne. The couple ruled well and are credited with forbidding humans to eat one another (implying a practice of cannibalism in pre-Egyptian civilizations?) and with teaching people many skills. Osiris taught humans about growing grain and making beer. Isis developed the practice of embalming, a skill she had to learn, according to legene, wnen she restored the multilated body of Osiris. Perhaps you remember the Greek myth of Demeter's daughter Persephone who was abducted by Pluto, the god of the underworld. Persephone was repelled by Pluto and vowed she would not eat until he released her.

Demeter roamed the earth searching for the daughter Persephone. During this periods of searching she met many mortals along with the family of Metanira who befriend her. Metanira's son Triptolemus was dying, and Demeter, to express her appreciation for the kindness shown to her, tried to make the boy immortal. Metanira, who did not realize that Demeter was a goddess, snatched her son from the ashes where Demeter had placed him as part of his transformation ceremony. While the boys's chance of becoming immortal was thwarted, Demeter nevertheles provided him with the knowledge of plowing and cultivating so that he could teach other men.

As she continued her search, Demeter found evidence that her daughter had been drowned, and she cursed the earth so that animals died and plants would not grow. When she finally learned the truth, Demeter pleaded with Zeus to restore her daughter to her. Zeus agreed to do so only if Persephone had eaten nothing during her stay in the underworld. Unfortunately, Persephone had finally given in, accepted a pomegranate offered to her, and eaten the pulp from one seed.

As a result, Persephone was forced to marry Pluto, but she had to live with him for only a third of the year, because of the seed she had eaten. This story is part of the Greek agricultural saga that explains the origin of agriculture and gives a rational for the 4 months of inhospitable weather for growing crops and the 8 month growing season which were linked to thetimes Persephone was below or above the ground. In Chinese mytohlogy, Shen Nung, the fictitious second emperor of China taught humans to use the hoe and the plow. Fire was his symbol, and the Chinese legend also attributes to him the knowledge of how to clear and burn forests.

In Central America, the Aztec god Quetzalcoatl appears in many myths. In one of the traditional story, he disguised himself as a black ant and carried a grain from Tonacatepel to Tomanchan and gave it to human beings for cultivation. To the south, along the western highlands of the central Andes, the great Inca empire developed its own gods and myths. For the people of this empire, the knowledge of how to sow seed came directly from Father Sun who sent a son and a daughter to teach people to revere him, build houses, and plant seeds. Needless to say, almost all the New World Indian groups as well as various tribes and cultures of Africa developed their various explanations of how people learned about the purposeful cultivation of plants.

The exception to the most common central idea of most of these legands, that the knowledge of agriculture was a "gift" from the gods, is found in the book of Genesis. In contrast to most legends, the Judeo-Christian account states that God said to Adam after Eve had eaten the forbidden fruit, "In the sweat of thy face

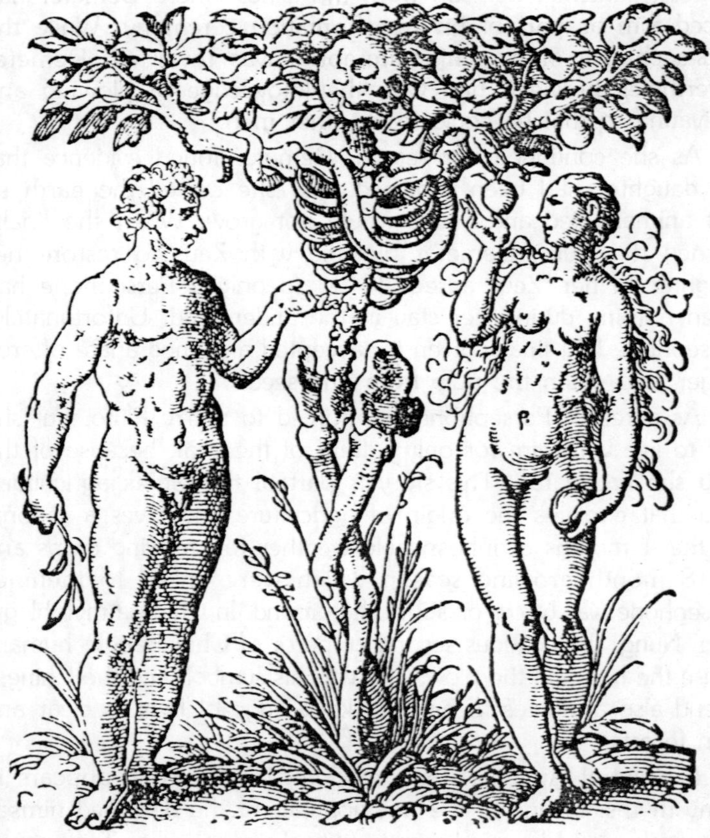

Fig. 1.1. In the biblical story of the origin of agriculture, Adam and Eve were driven from the Garden of Eden as punishment for eating the forbidden apple and from that time onward had to earn their sustenance by practicing agriculture.

shalt thou eat bread, till thou return to the ground." As punishment, "the Lord God sent him from the Garden of Eden to till the ground form whence he was taken." Obviously, in this case, having to practice agriculture was considered a burden incurred by misbehaving rather than a gift or blessing. In recent years, authors have rejected mystical or religious explanations for the origin of agriculture and sometimes replaced them with a "eureka! I've discovered it" theory.

In James D. Michener's book *The Source* this idea is perpetuated in the recounting of an innovative woman who hit upon the idea of purposefully planting seeds close to home. This eliminated the need to search for patches of grain. The logical result of this discovery was the subsequent spread of the knowledge to other human populations.

Edgar Anderson was a student of interactions between plants and animals, provided a somewhat similar idea in his famous "dump heap hypothesis". He conceived of the knowledge of cultivation as a consequence of the human practice of dumping refuse in a particular spot near the cluster of dwellings (once people began to live in some sort of settlement). Humans might have thrown on it seeds of grains, squashes, and other plants gathered from the wild. Since the wild relatives of many cultivated plants naturally occur in open and often disturbed habitats, the seeds would have reqdily germinated.

Soon, the dump heap would have become a thriving, if untidy, vegetable garden. An enterprising individual would have eventually seen that she or he could purposefully create a dump heap and thereby assure a supply of edible plant products close to home. Implict in this hyphothesis is the same idea as that of the eureka school, namely, that something happened which intitiated the discovery of how to grow plants. Besides eureka theoreies, there are larger number of considerable body of evidence indicating that nonagricultural people have the knowledge necessary to grow crops, they simply choose not to do so. By extrapolation, many botanists think that before the adoption of agriculture, people also knew how to grow plants. It is assumed that in ancient tribes, curers or shamen even grew a few plants that are useful for medicine. We might therefore ask, what tripped the balance? What caused humans seriously to begin the cultivation of crops if they had had the knowledge to do so for some time?

Why Farm?

Perhaps the most widely debated question about the origin of agriculture, and certainly one of the most interesting, is why humans, at more or less the same time around the world, suddenly adopted cultiation as a way of life. You might say that theanswer seems relatively simple. Practice of agriculture was just part of the progress of civilization from primitive cultures to our modern advance and technological society. This kind of philosophy about human determinism was a common notion in the late nineteenth, and even early twentieth, centures. Primitive humans were conceived of as living a deprived life. The progress of civilization and the improvement of the human lot were viewed as a series of revolutions, each of which led to a better stage in cultural evolution. However, within the last 20 years, these ideas have completely changed. Recent advances in archaeology, better dating of fossil material, and in-depth studies of modern nonagricultural peoples have indicated that agriculture was not suddenly discovered. The modern ideas and studies have suggested that primitive people are not deprived and the primitive agriculture does not improve such longer number of human. Instead, soem workers propose that people turned to agriculture only when forced to do so.

Obviously, these more recent theories conceive of agricultural practices as neither a technological breakthrough nor as an inevitable part of the human march to higher levels of civilization. In contrast, they view the adoption of agriculture as a logical step for people who know how to grow plants to take when it becomes necessary or the most practical course of action. If one assuems that humans did turn to agriculture when it best served their ends, the question immediately comes to mind, what ends? There have been two kinds of hypotheses, one emphasizing that agriculture was adopted when it best suplemented a change in life style and the second claiming that some form of stress must have prompted the change in food procurement habits. According to Carl Sauer human begun to cultivate crops because it served their needs.

Sauer envisioned that people had become sedentary before turning to agriculture and that once settled, collecting wild plant food at greater and greater distance from the settlements became so impractical that they began to raise a permanent supply of plant food nearby. According to Sauer, therefore, the people who turned to agriculture were not deprived, but settled and pensive, with enough leisure time to conceive of and implement ways of improving life. In Sauer's thinking, like Darwin's before him, people must have become sedentary before they adopted cultivation. Many other geographers, botanists, and anthropologists have searched for some event that forced humans to turn to agriculture.

One of the most cogently argued of such forces has been climatic change. About 50 years ago, it was proposed that the climate around the eastern Mediterranean Sea became drier following the retreat of the last major Pleistocene glaciation in Europe, forcing people who inhabited the Near East to live crowded together around wate sources. Within these oases, the theory states, humans had to practice agriculture in order to obtain enough food for survival. contrary to this theory, later evidence suggested that the climate around the Mediterranean became wetter, rather than drier, at the end of the Pleistocene, between 10,000 and 11,000 years ago. This increase in humidity would have favoured the extension of wild edible grasses into the Near East.

An increase in the abundance of wild grains might explain the greater use of them, but it does not explain why humans began to cultivate them. According to another explanation human started out gathering wild grasses, excess of them stored. Now this accumulated or stored grain provide stability to human. Human stay in one place to protect their hoard. Food surpluses could have spurred population increases until population sizes were too large to exist by simply gathering wild grains. Hypotheses that single out stress resulting from population pressure as the primary factor which caused humans to resort to agriculture have recently received much attention. Data taken to support a view that people turned to agriculture only when hunting and gathering could no longer provide enough food, come from the studies of modern nonagricultural people.

The assumption made here is, of course, that these modern hunter-gatherers are similar in many respects to ancient nonagricultural people. The most famous of these studies have been made by Lee of the !Kung bushmen of Africa (pygmies who live in the Kalahari Desert at the southern end of Africa). These studies indicate that, contrary to prior opinions, groups of hunter-gatherers do not lead lives of hardship or suffer from malnutrition.

Kung are very much selective in the plants that they use and the combination of all collected plants and few items of game provided adequate diete. In this way Kung are able to recognize about 105 species of edible plants use grams (g) of protein and 2,355 calories (cal) per day.

Estimates on modern nutrition lists often indicate that for a 154 lb male, 40 g of protein and about 2,700 cal per day is an adequate diet. In view of the fact that the !Kung are pygmies, their diet would thus seem to be very good. If we consider another method of comparison, the amount of work expanded to receive this amount of protein and caloric intake, we find that !Kung do not have to work harderthan primitive agriculturalists. The !Kung average about 2.5 days per week per person for food acquisition, or between 400 and 1,000 h per person per year.

In primitive agricultural societies, 1,000 person-hors per year is the lower limit. Obviously, a switch to agriculture does not provide a life with more leisure time, as had previously been supposed. When queried about plants, 'Lee and his associates concluded that the Kung possessed deep knowledge about many aspects of plant growth and ecology. If asked why they did not grow plants purposefully, the !Kung answer tended to be, why? Why should they indeed. Perhaps you might say the !Kung and such modern nonagricultural people live in especially productive areas, but since many of these groups live in the Kalahari Desert, this seems unlikely.

A more pertinent question is, if populations of ancient peoples increased to such levels that they were forced to turn to agriculture, why not the !Kung and similar tribes? As yet, this question has not been satisgactorily answewd, although recent evidence indicates that population levels are maintaned by extremely frequent nursing which plays a role in preventing ovidence. In fact, the weakest point in the argument of the poppnents of population pressure as the stimulus for turing to agriculture is that there is no direct evidence that suggests that population pressure stimulate the agriculture.

Circumstantial evidence, such as changes in the composition of the siet, as indicated from the analysis of the contents of ancient refuse heals, has suggested that there was a tendency toward less "preferred" food—mollusks or snails rather than large vertebrates, at about the tine of the emergence of agriculture. Obviously, it wi]l be very difficult ever to know with certainty if populations 10,000 yers ago were too dense in certain areas to be able to lice by means of a huntin and fatjering existence. As we can see, there is no consensus avout why humans becane farmers. As jack Harlan, one of the most respected contemporary researchers workinf on this question, has pointed out, there is no reason to assume that there is only one reason or that the same reason prevailed in different areas. People and populations under the same conditions may or may not respond in similar ways. Evidence reflects that agriculture arose independently in several parts of the world, it is likely that somewhat different pressures or inducements led pepople in diverse areas to begin purposefully sowing, harvestiong, and resowing plants in relatively permanent settlements.

Origins of Particular Crops

There are only few definite answers to our questions about the when, where, how and why of the origin of agriculture, there is better evidence for possible times and sites of the domestication of many specific crops. We seldom think about the fact that the variety of foods that we commonly eat have been drawn from numerous and widely separated parts of the world, but even typical Italian food such as tomato sauce, chicken cacciatore, and pizza would not have existed before the discovery of America because there were no tomatoes, no red and green peppers, and no zucchini in Italy. Italians obviously had a diet quite different 500 years ago from that they enjoy today. Similarly, what seems more American than apple pie or green peas, both of which were unknown in the New World until Europeans introduced them.

Fig. 1.2. Barley, an important item of trade in ancient times, as depicted on a Greek coin.

It is fascinating to trace the origins of our cultivated plants, some of their ancestral homes becomes similar to solving a mystery. Furthermore, as we become more aware of the need for maintaining diversity in our cultivated crops, we realize that such studies are not simply an intriguing pastime, but a necessity. Swiss

Fig. 1.3. Plants of Mediterranean or Near East origin. 1–sage, 2–thyme, 3–oregano, 4–parsley, 5–mint, 6–wheat, 7–barley, 8–oots, 9–fig, 10–lentils, 11–pomegranate, 12–olives, 13–leeks, 14–rosemary, 15–lettuce, 16–ortichoke.

was the first person who seriously undertake a study of plants and their origin, while a book titled (in translation), '*Origin of Cultivated plants*' was published by Alphonse de Candolle in 1982. In tracing the original homes of the cultivated species that interested him, de Candolle synthesized information gathered from

Fig. 1.4. Plants of New World origin. 1–Jerusalem artichoke, 2–pinto beans, 3–tomatoes, 4–pepper, 5–vanilla, 6–papaya, 7–avocado, 8–wild rice, 9–potatoes, 10–green beans, 11–peanuts, 12–corn, 13–jicama, 14–pineapple.

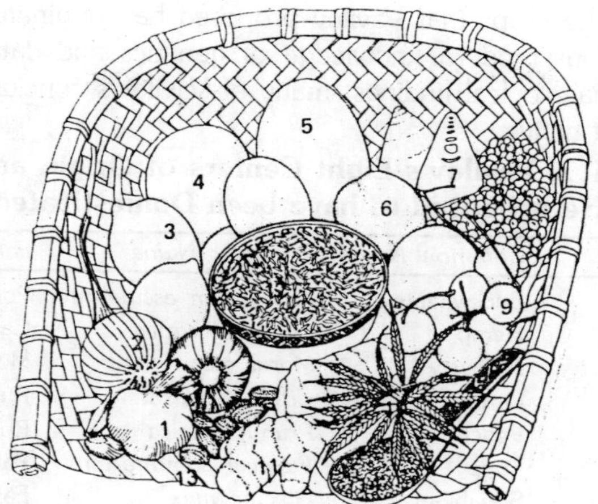

Fig. 1.5. Plants of Asian origin. 1–garlic, 2–onions, 3–lemons, 4–oranges, 5–mangoes, 6–peach, 7–bamboo shoots, 8–soyabeans, 9–cherries, 10–hemp, 11–ginger, 12–rice, 13–almonds, 14–tea.

studies of geography, linguistics, archaeology, and written history, which he combined with his extraordinary knowledge of plants of the world.

While his notions about how wild plants were transformed into cultivated crops seem quaint in view of our modern knowledge of genetics and selection, many of his conclusions about places of origin of domesticated have been shown to be correct. The next comprehensive treatment of crop plant origins came from a Russain about 70 years later Nikolay Vavilov tried to find centers of origin of cultivated species. Vavilov formed two assumptions. First, the s sites of original domestication are those areas where relatives of cultivated species can now be found. Second, centers should be areas in which one could find great amounts of natural variation in crops that are grown. From his research, Vavilow first delimited six, soon expanded to eight, major centers of domestication in the world.

Later, he and his collaborators increased the list still further, but the eight centers are the ones which are generally used. These centers are the Chinese center, the Indochina-Indonesian center, the Mid Eastern center, the Indian center, the Mediterranean center, the Abyssinian center, the formulation of six centers, the first two were considered as one, the southeast Asian center, and the second two as the southewest Asian center. Table 1.1. lists some of the crops that Vavilov proposed had originated in these centers. Our modern knowledge of genetics and data available from archaeological findings, many of Vavilov's conclusions seem somewhat native.

Table 1.1. Vavilov's Eight Centers of Origin and Some Crops He proposed to have been Domesticated in each

Center	*Common Name*	*Scientific Name*	*Family*
Chinese	Buckwheat	*Fagopyrum esculentum*	Polygonaceae
	Hemp	*Cannabis sativa*	Cannabaceae
	Mulberry	*Morus alba*	Moraceae
	Orange	*Citrus sinensis*	Rutaceae
	Peach	*Prunus persica*	Rosaceae
	Poppy	*Papaver somniferum*	Papaveraceae
	Soyabean	*Glycine max*	Fabaceae
	Tea	*Camellia sinensis*	Theaceae
	Tung	*Aleurites* spp.	Euphorbiaceae
Indian, including	Banana	*Musa* spp.	Musaceae
Indo-Malayan	Breadfruit	*Artocarpus altilis*	Moraceae
	Chickpea	*Cicer arietinum*	Fabaceae
	Citron	*Citus medica*	Rutaceae
	Coconut	*Cocos nucifera*	Arecaceae
	Mango	*Mangifera indica*	Anacardiaceae
	Peppper, black	*Piper nigrum*	Piperaceae
	Rice	*Oryza sativa*	Poaceae
	Safflower	*Carthamus tinctorius*	Asteraceae
	Sesame	*Sesamum indicum*	Pedaliaceae
	Sugar	*Saccharum officinarum*	Poaceae
	Yam	*Dioscorea alata*	Dioscoreaceae
Central Asian	Apple	*Malus pumila*	Rosaceae
(includes NW	Carrot	*Daucus carota*	Apiaceae
India)	Grape	*Vitis vinifera*	Vitaceae
	Onion	*Allium.cepa*	Liliaceae
	Pea	*Pisum sativum*	Fabaceae
	Pear	*Pyrus communis*	Rosaceae
	Radish	*Raphoanus sativus*	Brassicaceae
	Spinach	*Spinacia oleracea*	Polygonaceae
Near Eastern	Alfalfa	*Medicago sativa*	Fabaceae
	Fig	*Ficus carica*	Moraceae
	Flax	*Linum usitatissiumum*	Linaceae
	Hazelnut	*Corylus avellana*	Betulaceae
	Lentil	*Lens culinaris*	Fabaceae
	Melon	*Cucumis melo*	Cucurbitaceae
	Oats	*Avena sativa*	Poaceae

	Quince	*Cydonia oblonga*	Rosaceae
	Rye	*Secale cereale*	Poaceae
	Wheat, einkorn	*Triticum monococcum*	Poaceae
	emmer	*T. turgidum*	
Mediterranean	Asparagus	*Asparagus officinalis*	Liliaceae
	Beet	*Beta vulgaris*	Chenopodiaceae
	Cabbage	*Crassica oleracea*	Brassicaceae
	Carob	*Ceratonia siliqua*	Fabaceae
	Lavender	*Lavendula angustifolia*	Lamiaceae
	Leek	*Allium ampeloprasum*	Lilaceae
	Lettuce	*Lactuca sativa*	Astraceae
	Olive	*Olea europea*	Oleaceae
Abyssinian	Barley	*Hordeum vulgare*	Poaceae
	Castor bean	*Ricinus communis*	Euphorbiaceae
	Coffee	*Coffea arabica*	Rubiaceae
	Millet, African	*Eleusine coracana*	Poaceae
	pearl	*Pennisetum americanum*	Poaceae
	Okara	*Abelmoschus esculentus*	Malvaceae
	Sorghum	*Sorghum bicolor*	Poaceae
Mexican-Central American	Avocado	*Persea americiana*	Lauraceae
	Bean, common	*Phaseolus vultaris*	Fabaceae
	Cacao	*Theobroma cacao*	Sterculiaceae
	Corn	*Zea mays*	Poaceae
	Cotton	*Gossypium hirsutum*	Malvaceae
	Potato, sweet	*Ipomoea batatas*	Convolvulaceae
	Pepper, red	*Capsicum* spp.	Solanaceae
	Squash, winter	*Cucurbita moschata*	Cucurbitaceae
Central Andean (including most of South America	Manoic	*Manihot esculentum*	Euphorbiaceae
	Peanut	*Arachis hypogaea*	Fabaceae
	Pineapple	*Ananas* spp.	Bromeliaceae
	Potato, white	*Solanum tuberosum*	Solanaceae
	Pumpkin	*Cucurbita maxima*	Cucurbitaceae
	Rubber	*Heavea brasiliensis*	Euphorbiaceae
	Tobacco	*Nicotiana tabacum*	Solanaceae
	Tomato	*Lycopersicon esculentum*	Solanaceae

Current reconstructions of the process of domestication of a crop involve the compilation of information from cytoogy, ecology, plant systematics, archaeology, and ethnology. Careful assessment of the probable areas of origin of many of our important crops has now indicated that domestication occurred over a much more diffuse region than Vavilov supposed. Particularly in Africa and South America, there do not seem to be restricted regions in which notable preponderances of crops were were first domesticated. Jack Harlan has led in the synthesis of modern data and developed a concept of centers and noncenters. Most simply stated, noncenters are areas so large that designating them as centers is meaningless. A case in point is Africa. First of all the domestication of crop takes place in Africa then they were brought into cultivation in different parts of the continent. Designating the entier African continent as a center does not convery any information beyond the fact that a plant was probably native to Africa and first used by humans there. Designating Africa as a noncenter is a way of indicating the diffuse nature of the places in which various crops were domesticated.

According to Harlan the center of agricultural origins are Near East, north China, and Mesomerica, but in his interpretation these centers are areas in which agriculture arose independently. Each was, in his view, associated with a noncenter over which domestication of individual crops occurred. Nevertheless, while Harlan's views are probably very realistic considering the evidence for the independent origins of agriculture and much recent data about the areas of first cultivation of many important crop plants, workers still refer to Vavilov's centers when they discuss economically important plants. In this text, however, we will try to assess each plant independently rather than trying to force it into a system of centers of domestication.

2

Foods from Leaves, Stems, and Roots

Plants have been compared, in general terms, to animals turned inside out, with the leaves representing animal lungs, and the roots with all their ramifications, the intestinal system. This analogy of the Baron Justus von Liebig in his 1829 publication *National Law of Vegetation* reflected the view of some of the early investigators of plant physiology. While such quaint notions of the roots as the digestive organs of a plant seem ridiculous to us today, they are not much more erroneous than the common modern impression that roots are any underground portion of a plant and that all of the above-ground green or woody portions are stems or leaves. In this chapter we discuss each of these organs in more detail and relate their diverse natural structures and functions to the uses we make of them as food crops.

Biennial and Annual Stem, Leaf and Root Crops

Turnips and their Relatives

The array of vegetables produced by modifications of the shoots and roots of members of the Brassicaceae or mustard family exceeds that of any other plant group. The food crops of the Brassicaceae, including mustard and rapeseed oil, share a pungent flavour imparted by a class of compounds known as the mustard oil glycosides, or glucosinolates. While not humans in the quantities in which we consumer them, mustard oils have been shown to toxic to insects. Genus *Brassia* of the mustard family is the most commonly consumed vegetable in America. These are grouped together as the cole crops.

Interestingly, almost all of the leafy vegetables, in this genus belong to one species, *Brassica oleracea*. Each of these vegetables, kale, Brussels sprouts, cabbage kohlrabi, broccoli, and cauliflower, is produced by a different modification of the leaf or shoot system, each has its own history of domestication. Historical documentation shows that *Brassica oleracea* was cultivated by Greek as early as 650 B.C., although fossil evidence is lacking. By the time of the early Greek writings, there is menton of a type of leafy vegetable with a head which might have been cabbage, and another which sounds from the description rather like kohlrabi. Selection within the species was already advanced at an early date, and the train of selection of various modern morphological forms was underway.

Presumably the plants first taken into cultivation were rather loose-leaved, rank herbs. Forage kales [the European black cabbage (*Brassica oleracea acephala*)] are modern cultivated form resembling wild ancestral types. Due to their adaptation to coastal habitats, they were frequently exposed to salt spray. The waxy layer that protected these plants form salt damage is still evident in our modern cole crops and contributes to their resistance to drought and cold. The common cabbage (*Brassia oleracea* var. *capitata*) includes red and green cabbages with tight heads and smooth leaves.

A distinct variety, found less in the United States than the more compact type, Savoy cabbage (*B. oleracea* var. *bullata*) has looser and crumpled leaves. Although a headed cabbage was described by ancient Greeks (e.g., Pliny the Elder, 23 to 79 A.D,), the modern variety (*B. oleracea* var. *capitata*) is believed to have been developed in Germany where both red and white (= green) cabbages were grown by 1160 A.D.

Headed cabbages are formed when the terminal meristem of the primary shoot does not elongate and the "inner" leaves do not expand. The terminal bud remains buried deep inside the closely appressed leaves. European peasants have always given importance to cabbage since they would be harvested late in the fall and stored for considerable periods of time. Also for preservation through the winter, the shredded cabbage leaves could be packed and salted into earthenwere crocks to make sauerkraut. This process was in use since about 200 B.C. However the cold

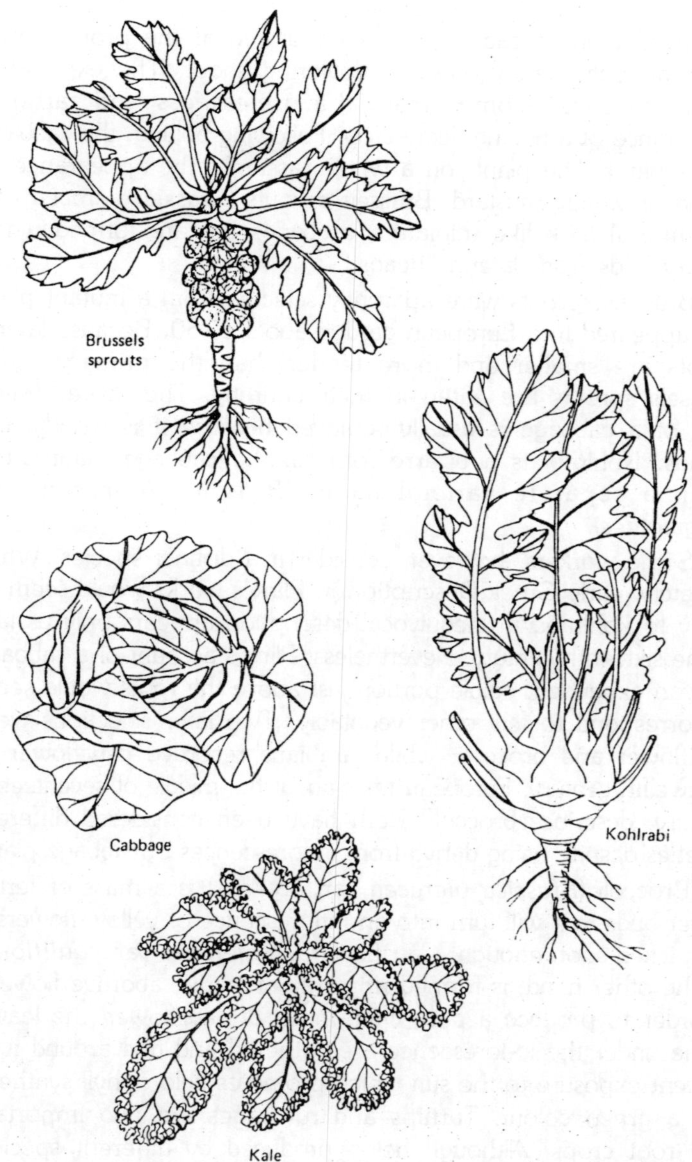

Fig. 2.1. The postulated proliferation of the varieties of edible brassicas.

tolerance that has made cole crops so popular in the north temperate zone has also dictated their limits of successful cultivation. These hardy vegetables should be planted in the early spring or fall because they mature best under cool temperatures.

In addition, if cabbages are not picked at the proper time, they can bolt, i.e., produce a flowering shoot. The expression "gone to seed" comes from the uselessness and tattered appearance of a bolting, leafy crop. Following bolting, the cabbage leaves bitter. The plant, on a whole, assumes the appearance of a wild or weedy mustard. Brussels sprouts (*Brassia oleracea* var. *gemmifera*) look like miniature cabbages but are formed when axillary buds from lateral "heads."

Brussels sprouts were aparently selected from a mutant plant that appeared in a European garden about 1750. Because lateral shoots are smaller and more tender than the primary stem, Brussels sprouts are eaten in their entiretly. The "core" (main axis) of a cabbage is usually removed because it is woody and fibrous. Kohlrabi is a bizarre form of the cabbage plant often given a separate varietal name (*Brassica oleracea* var. *gongyloides*).

Some workers, however, consider it a distinct species. While a sketchy early Greek Description in Pliny's works would seem to apply to kohlrabi, a unequivocal description appears only as late as the sixteenth century. Nevertheless, Pliny's account of a cabbage with an expended, edible portion just above the roots would seem to correspond to few other vegetably. Two later selections were cauliflower and broccoli, while Arabians recorded cauliflower in the twelfth century, European texts about the middle of seventeenth century, describes broccoli. Both have been considered different varieties despite being derivd from inflorescences of cabbage plant.

Broccoli (*Brassica oleracea* var. *botrytis*) is a mass of fertile flower buds that will turn into an inflorescence of yellow flowers if not picked soon enough. Cauliflower (*B. oleracea* var. *cauliflora*), on the other hand, is formed primarily of sterile, abortive flowers. In order to produce a pure white head of cauliflower, the leaves borne under the inflorescence are gathered and tied around it to prevent exposure to the sun which promotes chlorophyll synthesis and a green colour. Turnips and rutabages are two important cole root crops. Although being produced by different species, their striking similarity to each other after harvestation, have led to their confusion even at the produce counter in grocery stores.

The parts of turnips and rutabagas we eat actually contain more than just the taproots of the plants. The upper part of the globular structures is the hypocotyl, or the portion of the plant

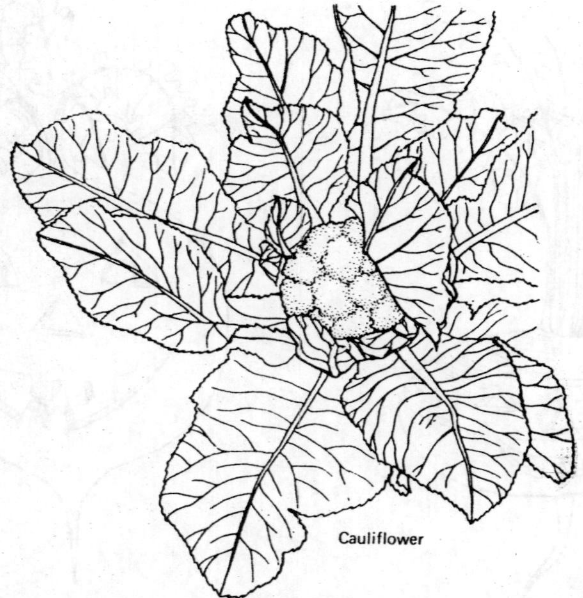

Fig. 2.2. Cauliflower.

where the stem and root intergrade. Most of the mass is root tissue, however, and the vegetables are more properly considered roots then stems. Turnips, or turnip rapes, are the roots of *Brassia campestris*, thought to have grown originally wild in Europe and Asia. In the Indian writings of 2000 B.C. there are found references to a plant with similar appearances to turnip. Only in the thirteenth century, did English cultivation of turnip begun.

Defferent cultivars of this species yield Chinese cabbage. The English name *turnip* comes form the same etymological base as the verb "to turn" because turnips are so smooth and perfectly formed they appear to have been turned on a lathe. The flesh of the turnips sold as vegetables is usually white. The roots themselves are flat on the top and generally tinged with purple. Yellow-fleshed varieties are grown but are not commonly found in stores. Turnips have for some reason always been held in low esteem. In Roman times they were spoken of in derogatory terms, and they because they were a favourite item to throw at miscreants.

The Aryans disdained them because they were eaten by Indian races. Young German women in some areas would present suitors that they wanted to reject with a plate of boiled turnip. They

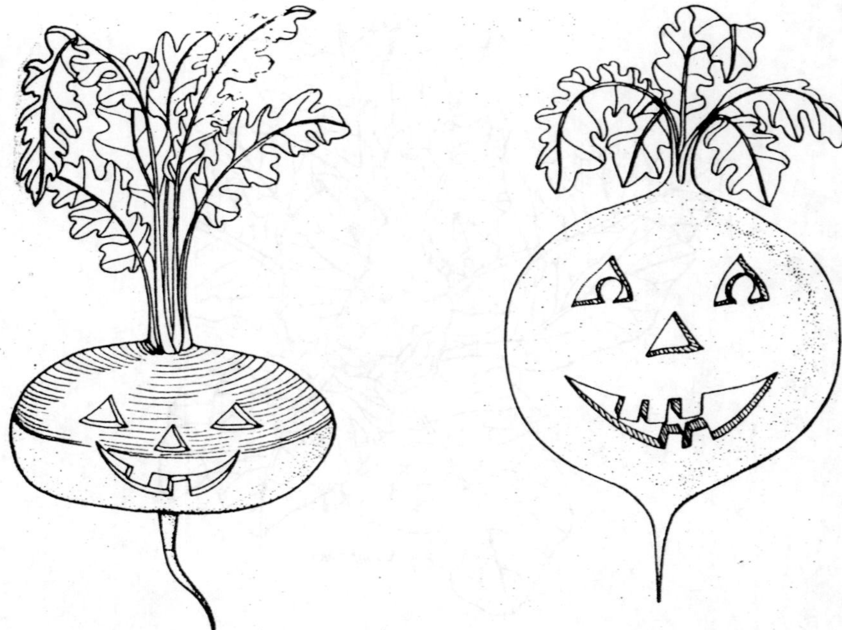

Fig. 2.3. Turnip (left) and Rutabagas (right).

majority of Europeans, however, consumed great quantities of turnips throughout the middle Ages and still include them as a part of many winter meals. Yet this dependency has not seemed to change the opinions of most people about the vegetable. Even in the United States, children are led to believe that nothing could be worse than having to eat turnips.

Most people believe that they have a strong flavour, but they have perhaps never had a good turnip or have mistakenly confused the vegetable they ate with its relative, the rutabage. Rutabages, or Swedish turnips, come from *Brassia napus* and are more nutritious, but have a more pronounced flavour, than their smaller cousins.

Rape, or colza oil are produced from an oil seed crop, which have been selected from the cultivars of Rutabaga species. Barrungthii, the history cultivation of Rutabaga has parallel that of turnips. Since their chromosome number and complement is double that of cabbages and turnips, it can be inferred that Rutabagas have resulted from hybridization and polypoloidization of cabbages and turnips. In Europe, rutabagas are often used as food for domesticated animals. In U.S. markets, rutabagas are yellow and larger than turnips. The tops are cut off before marketing, obscuring the fact that, unlike turnips, they are pointed at the upper end. Shippers often coat rutabagas with wax to help prevent drying.

Radishes (*Raphanus sativus*, Brassicaceae) are an important root vegetable on a worldwide basis, but in the United States, they have been relegated to the position of a garnish, carved to resemble roses and to be observed, not eater on relish trays and patters. Other cultures hold the peppery roots in much higher esteem. Inscriptions from Egyptian tombs dated to be almost 4,000 years old show that radishes were important in this ancient civilization. Egyptians not only ate the roots, but also valued radish seed oil. In China, selected cultivars are still grown as an oil seed crop. In Japan and many other oriental countries, radishes are among the most important vegetable.

In the orient, however, the roots are not the small red spheres that are consumed in the United States, but rather, elongate white, or giant black-skinned forms known as *daikons*. Although the ancestor of the domesticated radish is not known with certainty, the large straight-rooted typed are probably similar to the wild types form which modern variants have been selected. The globular form was developed only in the eighteenth century ad initially white. The bright red colour was the result of a later mutation that was perpetuated by man.

Lettuce and its Relatives

Lettuce, chicory, and endives belong to the Asteraceae, one of the largest angiosperm families, containing over 13,000 species. The member of this family are easily recognized by the characteristic heads of many small flowers. Yet, despite its large number of species, the family has not contributed many plants to the human diet. There are, however, a few notable exceptions. The most plants to the human diet. There are, however, a few notable exceptions. The most important of these is lettuce (*Lactuce sativa*). In ancient times, it was thought that the milky juice exuded from a cut lettuce possessed medicinal virtues similar to those of opium latex. Because of the milklike sap, Linnaeus chose the generic name *Lactuca* derived from Latin *lac*, or milk. Some forms of lettuce may be have been cultivated as early as 4500 B.C. and lettuce on Egyptian tombs. The Romans later enjoyed various greens tossed with olive oil and vinegar as preludes to their renowned feasts.

Like many domesticated species, there is no wild plant that can be definitely associated with lettuce, thought *L. serriola* and some other species of the Mediterranean region appear to be very close relatives. The divergent selection to which lettuce has been subjected can be seen throughout history. It has particularly evident by the 14th century, when salads became a rage in England, the housewives had at their disposal, over 50 different type of greens, many of them being varieties of lettuce. A virtuous homemaker homemaker arranged her assortment of buttery, crisp, sweet, and bitter greens with snippings of herbs, violets, and other flowers so as to give them an aesthetic appeal as well as an interplay of tastes. In contrast, the salad greens in our modern grocery stores seem very uninteresting. The lettuce types we do encounter today can be categorized into heading loose-leaf, and cos types.

Because of its tight, round heads, Iceberg lettuce—a heading type, has become the most common type sold in U.S. ship well. Other forms include loose-leaf types such as salad-bowl, red-tipped, and oak-leaf lettuces. The primary cos type with stiff, rather elongate leaves, is romaine. Like cabbage, lettuce heads are formed by the suppression of the terminal bud, and they will also bolt if not picked soon enough. Flowering lettuce is a rank-looking plant about I-m tall with small yellow heads resembling those of dandelions. Belonging to a related genus *Cichorium*, not *Lactuca* are endives and chicory which have graced salad bowls as long as lettuce.

In the United States, the salad delicacies known as Belgian endives come from the perennial species *Cichorium intybus*. Their production begins in spring, when small tender head seeds are sown and then until the first frost the plants are allowed to grow freely, after which they are dug and tapped. The thick root stocks are then planted in sand and blanched (forced to initiated new shoots in the absence of light). Under these conditions, chlorphyll synthesis is prevented or greatly reduced, and mild-flavoured torpedo-shaped head are produced. In Europe, these forced, blanched heads are called *witloof*. If the plants are allowed to leaf naturally, they produce large, rather bitter leaves that are also used as salad greens. *Cichorium intybus* is known as chicory

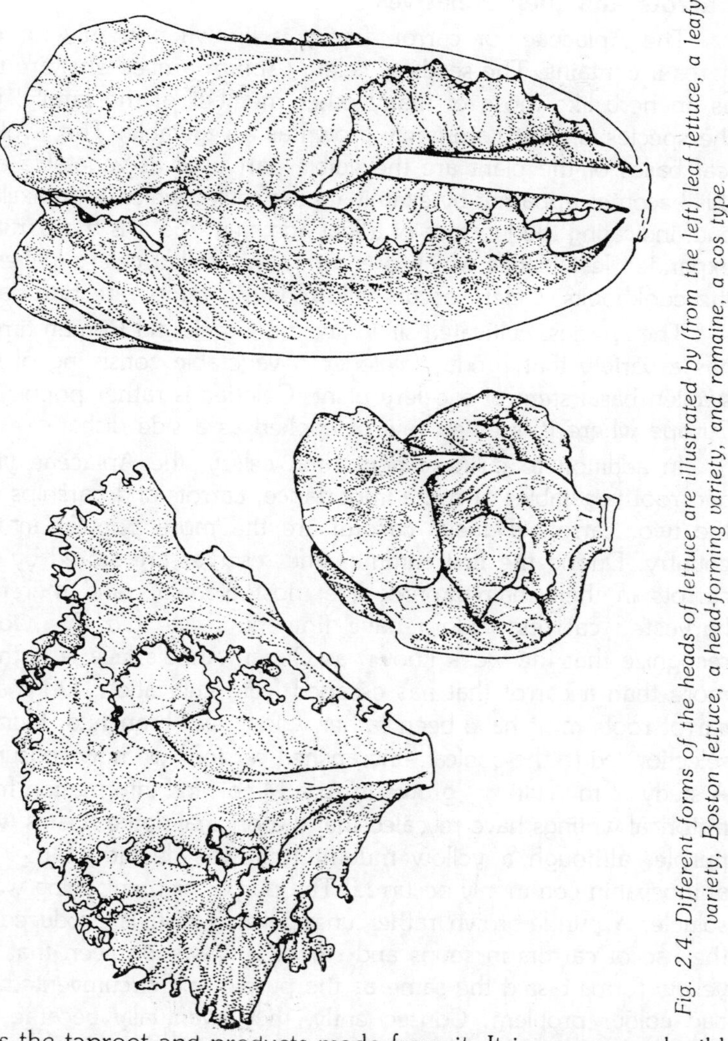

Fig. 2.4. Different forms of the heads of lettuce are illustrated by (from the left) leaf lettuce, a leafy variety, Boston lettuce, a head-forming variety, and romaine, a cos type.

as is the taproot and products made from it. It is an escaped wild plant in U.S. The roots are roasted and ground in Europe and parts of S. America. The powder is used mixed with coffee of as a coffee substitute. Adding to the confusion is the fact that in American supermarkets, the salad green known as chicory or escarole comes from another species, *C. endiva*. The heads of this annual species can also be blanched by tying the large, outer leaves around the inner "heart" of the head.

Carrots and their Relatives

The Apiaceae, or carrot, family is known primarily for the herbs it contains. The seeds of celery (*Apium graveolens*) are use as an herb like those of many other member of the family, but the species is most commonly grown as a vegetable. The swollen leaf bases of the plant are the edible part used as vegetable and this becomes apparent by examining the stalks, which is an axillary bud, indicating the juncture of a leaf and stem. So it is the partially expanded leaf blades which are commonly called celery leaves is the cookbooks. Celery occurs wild today in temperate Eurasia.

The species, cultivated since ancient Greek and Roman times, has a variety that produce celeriac a vegetable consisting of the swollen basal stem of a celery plant. Celeriac is rather popular in Europe where it is eaten raw of cooked as a side dish.

In addition to culinary herbs and celery, the Apiaceae yield two root vegetables of some importance, carrots and parsnips. Of the two, carrots (*Daucus carota*) are the more popular in this country. During the first summer, the reserves are stored by the carrots in their main taproot like most biennials and therefore harvested carrots seldom have flowers. If we did, we would recognize that the week known as Queen Anne's lace is nothing more than a carrot that has escaped from cultivation. Originally, carrot roots must have been rather woody and branched. Human selection led to the conical, unbranched versions we now consume. A study of the cultivars grown in Asia Minor and information from historical writings have revealed that the original domesticates were purple, although a yellow mutant that was devoid of the red anthocyanin commonly occurred. The purple pigment being water soluble, a purple-brown rather unappetizing dish is produced by the use of carrots in soups and stews. It was soon seen that the yellow forms tasted the same as the purple and circumvented the bad colour problem. Consequently, they eventually became the most commonly grown type. Latter selection intensified the colour. Carrots are now eaten not only cooked, but as a raw vegetable.

Carotene is the pigment responsible for the orange colour of carrots. Carotene chemically is a pigment consisting of two vitamin A molecules joined end to end. During digestion, carotene is immediately broken down to vitamin A by the mucus lining of the intestine and by the liver. Vitamin A, in turn, is almost identical to the chemical receptor retinal present in the rod cells of the retina of the human eye. Rod cells are important in vision because they gather light at low intensities. So apart from its various functions, vitamin A also functions for supplying the precursor for the important pigment of the eye's rod cells, thus proving to some extent the old wives tale that eating carrots can help you see in the dark. However, eating carrots merely helps to ensure that an individual receives enough vitamin A. Beyond a threshold, additional vitamin a cannot improve vision.

In appearance, there is a close resemblance between parsnips (*Pastinace sativa*) and carrots with the exception that the former are pale yellow instead of orange. Although being sweeter than carrots, they are not as popular in this country and also seen to fall into the category old fashioned vegetables. They seem to fall into the category of "old-fashioned" vegetables. Like carrots, they are native to the eastern region of the Mediterranean and were eaten throughout the Middle Ages in Europe. The pilgrims brought them to the New World where they became one of the few European crops enthusiastically adopted by North American Indian.

Beets

Beets (*Beta vulgaris*, Chenopodiaceae), although being considered by most of us as small round deep red roots that turn up as a side dish, pickled at salad bars, or as a primary ingredient of Russian borscht soup, isanomalously placed among the most important crops in the world. In addition, the species provides three other crop plants, the *mangel-wurzel*, Swiss chard, and the sugar beet. The first is a cultivar with a yellowish-white root that is used as a cattle feed. Chard is a leafy vegetable.

The sugar beet the most economically important of all of the products obtained from the species, is discussed later along with sugar cane. It is believed that chard (*Beta vulgaris* subsp. *cicla*), or chardlike greens, has been eaten by humans since prehistoric times in the Mediterranean region where beets are native. Early Greeks mentioned varieties of chard in their writings. The roots of Swiss chard plants are stringy and more branched then those of the subspecies which yield edible beets and are consequently not generally eaten.

Spinach, an Annual Leafy Crop

Spinach (*spinacia oleracea*), like beets, is a member of the goosefoot family (Chenopodiaceae). In salads, lettuce is often substituted by Spinesh now, although the Romans could not have include it among the salad green, because, the species being native to western Asia, appears to have been domesticated after the fall of Roman Empire. The cartoon character Popeye the Sailorman was originally devised to cajole children into eating spinach because of the large amounts of iron it contains. Unfortunately, it is now known that much of the iron in the vegetable cannot be absorbed because calcium oxalate in the leaves forms a chemical complex with the iron, hindering its absorption.

Perennial Green Vegetables: Artichokes and Asppracus

Perennial Habit

When farmers plant a perennial crop, they use procedures quite different from those followed by farmers cultivating annual or biennial plants. Great care is taken in the preparation of the soil for perennial crops because once the plants start to grow, they remain in place for many years. In order to allow the plants to became well established and to store up reserves in the root system, harvesting of crops from perennial vegetables is delayed until 2 or more years unlike the annual or biennial species. However, as the years progress, the farmer reaps increasingly larger harvests until plants begin to senesce. Generally after every 15-25 years, perennial temperaate plants are replanted.

Artichokes

Artichoke, one of our most distinctive vegetables, derives its name from the Arabic word *ardischauki*, meaning "earth thorn". Artichokes (*Cynara scolymus*) are members of the Asteraceae, but unlike lettuce and the other leafy family members, they are cultivated for their immature flowering heads. When we eat an artichoke, we eat the carbohydrates stored in the tender portions of the fleshy bracts surrounding the undeveloped flowers. The undeveloped flowers and the chaff intermingled with them form the inedible "choke" of the artichoke. The swollen part of the stem (receptacle) on which the choke borne is the savored artichoke 'heart'. If allowed to mature, an artichoke becomes a beautiful, large, blue thistle. In order to ensure uniformity and hasten plant establishment suckers or basal branches are generally used to plant artichokes.

Asparagus

Asparagus (*Asparagus officinalis*, Liliaceae) is a dioecious perennial native to the temperate scrub communities of southern

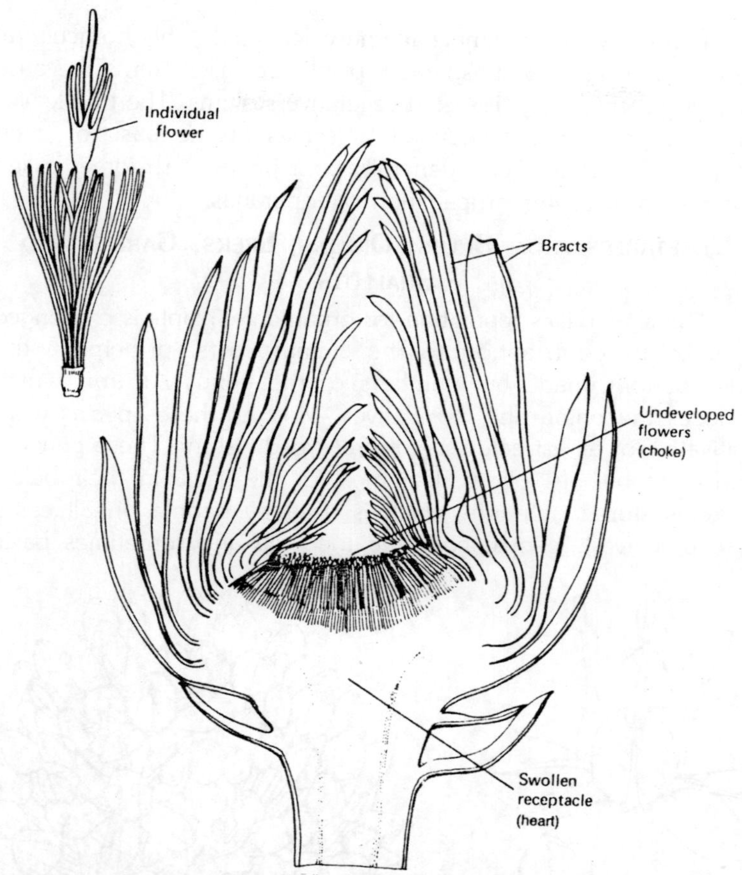

Fig. 2.5. An artichoke cut in half shows that the vegetable is really an immature flowering head.

Europe, western Asia, and northern Africa. These habitats are subjected to periodic burning, and due to presence of robust underground system of rhizomes. Many natively occuring plants of these regions have the ability to sporut rapidly. Humans have made use of this resprouting habit by growing plants in habitats where plants die back in the winter and send up new shoots in the spring. The vegetable consists of young, unexpanded shoots that are cut where they emerge from the ground.

Limited period of production occuring only in the spring and harvestation of only a certain portion of the new shoots without damaging the plants makes asparagus expensive. Shoots that are allowed to mature expand into ramified branch systems, or "cladodes," which resemble feathery leaves. In the horticultural trade, some varieties of asparagus plants are called "ferns" because of their green, highly dissected vegetative systems. The true leaves of an asparagus are reduced to scales at the base of each cladode. Since seed set is generally poor prices of rhizomes called crowns are used for propagation of asparagus.

Vegetables from Bulbs: Onions, Leeks, Garlic, and Shallots

The lily family's reputation for ornamental plants is challenged only by its contribution to the culinary arts, principally the contributions made by members of the genus *Allium*, which includes the onion and its relatives. Most of these species were cultivated, or at last eaten, before recorded history. These pungent herbs are bulbous plants, like most of the lilies. In their bulbs, there is stored nutrients which is used by people. In all cases (except chives), people consume the bulb and sometimes basal

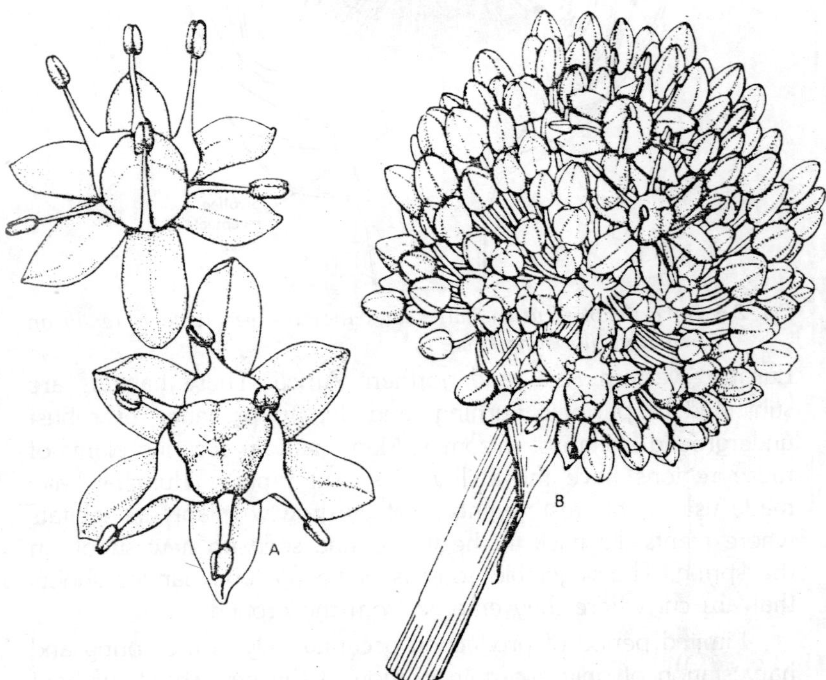

Fig. 2.6. The lily family provides us with the onion and its relatives. A—Onion flowers; B—Onion inflorescence.

portions of flattened leaf blades. Onions (*A. cepa*) and garlic (*A. sativum*) probably originated in central Asia and the leek (*A. ampeloprasum*) in the Near East. All were cultivated in Egypt, however, by the year 3200 B.C. Onions are grown for their single, large bulb, and garlic for its cluster of bulbs, each of which is called a clove. The bulbs produced by leeks are quite similar and are differentiated only slightly from the basal portions of leaf blades.

Consequently, a substantial portion of the leaf blades of leeks as well as the swollen basal portions are eaten. Another cultivated member of the genus, the chive (*A. schoenoprasum*), is eaten for the leaves alone. Medicinal properties have long been associated with almost all members of the *Allium* group. At one time or another, various species have been prescribed for bad eyesight (Hippocrates), voice improvement (Nero), baldness (Gerarde), or as a cure for the common cold, acne, arthritis, hypertension, or poor digestion. General ulysses S. Grant was so convinced that a supply of onions was needed to protect his troops from dysentery that a supply of onions was needed to protect his troops from dysentery that he declared "I will not move my army without onions."

With many health food enthusiasts advocating their use and others denying any therapeutic benefit, there is still a controversy

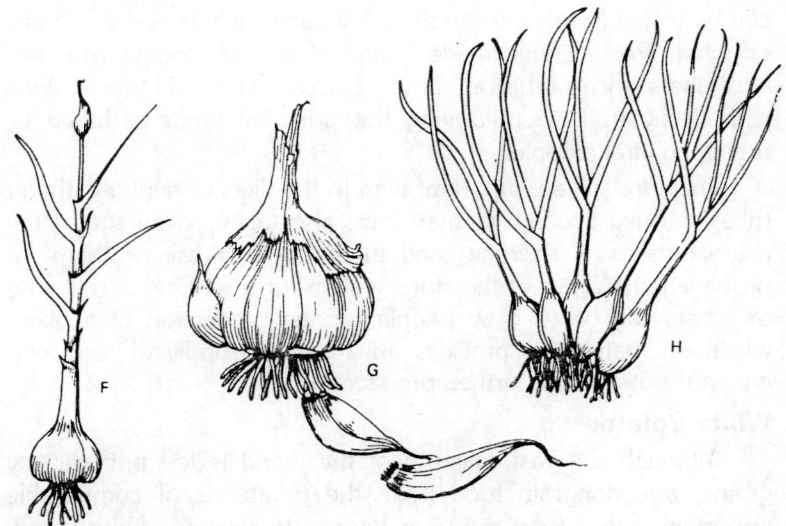

Fig. 2.7. Onion and its relatives: Garlic, garlic bulbs, and shallots.

regarding the real effect of onion and garlic. The pungent quality common to all of the alliums is linked with the compound that makes us cry when cutting onions. When the cells of an onion are ruptured, volatile sulphur compounds (including methyl di- and trisulphides and n-propyl di- and trisulphides) are released. When these compounds dissolve in the fluid covering our eyes, they from sulfuric acid. The tears brought on by the burning sensation only provide more water into which the compounds can dissolve. These effects can be diminished by submerging an onion in water while it is being sliced of by freezing the onion before slicing, since chemicals are water soluble.

Starches from Tubers, Rhizomes, and Roots

Tubers and Rhizomes

The important modification of stems has been for underground growth. Underground stems serve as overwintering organs and/or for asexual reproduction. Particularly adapted for human use are those underground stem that store nutrients. A seeding draws the nutrients it needs for initial growth from either the cotyledons or the endosperm. For a plant which reproduces vegetatively from underground stems, nutritive material can be derived by the newly emerging shoot, directly from the 'parent' plant's vascular system unless its own roots have been established, or reserve nutrients can be pulled from an underground storage such as fleshy rhizome or tuber. People have made abundant use of storage rhizomes and tubers since prehistoric times. Undoubtedly, wild tubers along with fleshy fruits were among the principal foods gathered by preagricultural people.

They are still an important item in the diets of hunter-gatherer tribes that exist today. Human have also taken advantage of the characteristics of rhizomes and tubers in their use or them for propagation. A part of the storage stem, or a few 'seed' rhizomes or tubers can be used by people for the production of a plant, which on maturation provides an abundant supply of food plus enough material for further propagation.

White Potatoes

Although the vast majority of the world is fed primarily by grains, one nongrain food item, the potato, is of comparable importance as a food staple in temperate regions of the world. Surprisingly, despite their modern importance, potatoes (*Solanum tuberosum*, Solanaceae) have been known outside of America for only a few hundred years. The commonly cultivated potato and several less well known tuber-bearing members of the same genus are natives of South America. Spanish explorers found potatoes being grown by Indians all along the Andes from Colombia to Chile.

By a primitive method of freeze-drying, Indians used to preserve potatoes in the high elevation areas of central Andes. A temperature fluctuation from below freezing point to almost 15°C (84°F) during the day each 24 hours can be observed in these areas. In addition, the climate is dry for most of the year. These combined climatic factors allow Indian to spread harvested potatoes on the ground and leave them overnight to freeze. The next day, the Indians stomp on them to squeeze out as much water as possible. A dry mass of cellulose and starch, known as *chuno* is yielded by this process after a few days. Since this method is not as rapid as modern freeze-drying (and ruptures the cells which is avoided in modern techniques), chuno often testes moldy, but once completely dry, it keeps for months. While the Europeans never favoured chuno, the potential of potato tubers as source of food was quickly spotted by the Spaniards. As a result they introduced these tubers into Spain and used them on return voyages to Europe. The rest of Europe learned about potatoes from an independent introduction to England.

Whether Sir Walter Raleigh or Sir Francis Drake first brought the potato to England is the subject of some debate, but evidence seems to favour the latter captain. A folktale says, that Queen Elizabeth I accused Sir Frances of trying to poison her when her cooks served a salad of potato leaves rather than a dish of cooked tubers from the plant delivered to them by Drake. Nevertheless, the potato was destined to become a great success in England and, soon, an greater success in Ireland. The enthusiastic reception received by potatoes in Ireland was related not only to its good growth in country's cool, but also to the fact that enough calories could be produced to meet the needs of an entire family just on a small amount of land.

In additions, Ireland was in the midst of a rebellious conflict with England and potatoes could be buried to hide them from enemy troops that periodically swept into Irish towns. By the 1840s, Ireland was not only fond of potatoes, she was completely dependent upon them. The entire country had essentially adopted a monoculture of the tuberous plants. A week's food for an average family of five typically consisted of about 40 lb of potatoes, several pounds of meat (probably mutton), a few loaves of bread, a few quarts of beer, bits of bitter, sugar, and tea. Potato blight, caused by a fungus *Phytophthora infestans* reached Europe in about 1845. There was large scale destruction of virtually all the Irish and British Potato crops within 5 years.

It is estimated that during these 5 years at least 1 million people (some say 2 million) died of starvation and over another million emigrated, many to the United States. After this disaster, special attention was paid by farmers to selection of disease resistance varieties. They also became aware of dangers of dependency on a single crop. Luckily, at about the same time, a fungicide (Bordeaux mixture) was developed which helped in controlling the blight. Today, potatoes are grown throughout temperate parts of the world and in upland tropical areas.

Planting is almost exclusively by seed tubers or by planting portions of the tubers which contain dormant buds. While we generally see only a few potato varieties in the United States, hundreds of varieties have been developed. Since the seeds are almost always nonfertile, pieces of tubers have been used for potato growth, but their use also creates problems because of presence pathogens, particularly nematodes and fungi, on these tubers. Gardeners can help reduce infestation in their crop by buying only certified "disease-free" seed tubers. Recently, a new variety of potato called "Pioneer" has been produced which can be grown from seed. The use of new variety has caused some excitement since use of seeds will not only help in reduction of pest problems but will also allow new breeding programs to be undertaken.

Yams

Yam is the second most important tuber in term of world production. Although the sweet potato (*Ipomoea batatas* Convolvulaveae) is often called a yam in the United States, true yams belong to the genus *Dioscorea* (Dioscoreaceae). There is some debate about whether a yam is derived from stem or root tissue; most authorities consider that the cultivated species have tubers. However, it is possible that some of the species have storage organs that are derived from rots rather than from stems.

Africa, Asia and S. America are home to different species of yams, and domestication appears to independently occured on each continent. The species traditionally cultivated in three native areas are *D. rotundata* and *D. cayenensis* in Africa, *D. alata* and *D. esculenta* in Asia, and *D. trifida* in the New World. Today, however, *D. rotundata* is grown throughout the world in the largest quantities. In all aspects of the culture, yams are important in many parts of Africa and Asia. Ritual yam cults are found in both New Guinea and Melanesia. Planting of ceremonial yams in special gardens are among the practices of these cults.

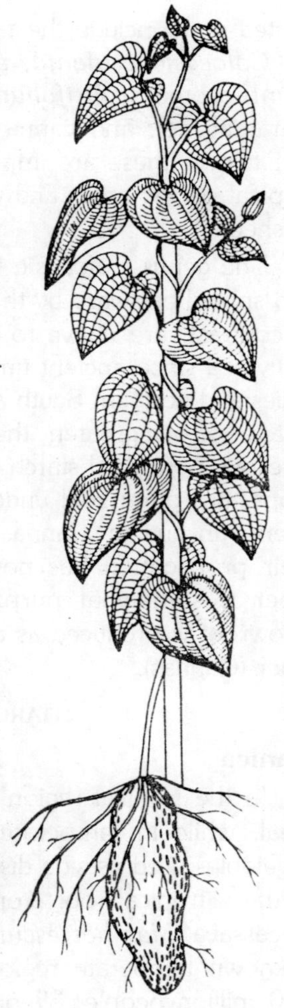
Fig. 2.8. Yam.

The size of these yams, which can reach up to 50 kg (122 lb), is thought to reflect the grower's status in the community. The ceremonial yams are used for gifts in ritualized exchange. Yams are held in such high esteem that only men are allowed to cultivate both eating and ceremonial yams. A yam festival is held at harvest time, during which tubers are dressed of covered by masks. Propagation of yams is yams is by asexual means using pieces of tubers with buds. Presence of oxalic acid (or oxalates) in cells just below the skin of the tubers is responsible for the poisonous or semipoisonous nature of many of the cultivted species of yams. The potential problems caused by crystals can be eliminated by peeling and boiling.

Yams are higher in protein than many of the other starchy by the crystals. Yams are higher in protein than many of the other starchy tuber crops, but unfortunately, yam cultivation has been replaced in many areas of the tropics by the of different tuberous crops which are easier to grow. Other tubers grown on a more limited scale include the taro, eddoe, dasheen, or "old" cocoyam (all *Colocasia esulenta*, Araceae), malangas, yautias or tannias (*Xanthosoma sagittifolium*, Araceae), arrowroot (*Maranta* spp., Marantaceae), and canna (*canna edulis*, Cannaceae). Only the first two of these are important crops plants. The staple food in tropical pacific areas and widely cultivated in west Africa also are Dasheens.

The genus *Colocasia* is native to east Asia and Polynesia but had spread eastward by the time of the Egyptian empire. Tannias, in contrast, are native to the New World where they have been cultivated since ancient times. Arrowroot, which is native in West Indies and northern South America, has a limited area of cultivation. Instead of being eaten, these tubers are used to make fine, easily digested, powdered starch. Adulterants, starch from other tuberous crops, are often sold under the name "arrowroot," but they are inferior in quality. Cannas are also native to South America, but their propagation has now spread throughout the tropics. The tuber is somewhat purplish. A starch known as Queensland arrowroot is produced as a result of drying and powdering of this tuber (Cannas).

STARCHY, ROOT CROPS

Manica

In the tropical region, major root crops constitute the entire meal, while in temperate regions, most of the root crops are vegetables used as side dishes during a meal. The most important of these tropical root crops has many names, manioc, manihot, or cassava (*Manihot esculenta*, Euphorbiaceae). While practically unknown temperate regions, manioc is the staple food of over 500 million people. 37 percent of the calorie consumed in Africa and 11 percent caloric intake in Latin America is contributed by it.

As is true of all the world's staple foods, manioc is almost pure starch. The storage roots are produced by monoecious perennial plants native to the tropical areas of the New World. There are hundreds of local varieties or land races of this species since they have been grown for so many thousands of years by small Indian tribes. The species is known as cultigen only. There are problems in interpretation of the evolutionary history of manioc and determining of the area of initial cultivation. This problem has been due to variability of the species and its widespread

Fig. 2.9. The manioc, or cassava, plant.

cultivation. It appears likely that it was domesticated independently in two regions, the semiarid areas of southern Mexico and adjacent Guatemala, and northeastern Brazil.

Archaeological evidence from mexico dates from only the first millennium B.C. Griddles for baking manioc bread dated to the year 2000 B.C. have been uncovered in South America. Still it is southern Mexico, where there is growth of wild species which appear to be most closely related to manioc. Manioc seems to be a very suitable crop for tropical region due to the following reasons:

(i) A good growth in both arid and wet climates.

(ii) Resistance to insect and fungal pests.

(iii) Production of storage root in poor soil.

(iv) Requirement of a minimum of agricultural labour a high yield per unit area.

Planing consists simply of clearing the ground and inserting into it 30-cm-long pieces of the stem which have several axillary buds. Each plant produces several storage roots which can be harvested in about 18 months or left in the ground until needed. Unfortunately, manioc has its drawbacks. Despite their high carbohydrate content (about 35 percent by weight), the roots contain very little protein (1 percent) and essentially no vitamins.

Moreover manioc roots can be eaten by humans only after the removal of a poisonous compound present in its roots. The poisonous compounds are two cyanogenic glycosides, linamarin and lotaustralin. These hydrolyze in the presence of the enzyme linamarin and lotaustralin. These hydrolyze in the presence of the enzyme linamarase to form, among other compounds, hydrogen cyanide. The enzyme and the glycosides come into contact if there is any bruising, cutting or grating of the roots. this leads to cyanide production. Roots that have little or none of the glycosides are known as sweet maniocs. Those with appreciable amount of the glycosides are called bitter maniocs. If present, the glycosides are located primarily in the outer cortical cells of sweet forms, whereas the toxic precursors are dispersed throughout the roots of bitter varieties. There are two major morphological types of manioc apart from the bitter and sweet forms. Smooth-skinned tubers and parts with silver-gray stems make up the first type whereas rough skinned tubers and red or purple-tinged stems make crop the second group.

Unfortunately, the morphological groups do not coincide with the groups based on cyanogenic glycoside content. In fact, in it is impossible to tell without prior knowledge of the kind of manioc that was planted whether it is of the bitter or sweet type. The fact that both forms are often processed similarly, particularly in south America, accounts for this lack of certainty. The roots are sold as dry, heavy, starchy lumps in Africa, where the sweet types were primarily introduced and are most extensively grown. These detoxified lumps can be ground into manioc flour to be used in breadlike products, or they can be reboiled to yield a rather gelatinous tasteless, starchy vegetable reminiscent of boiled potatoes. In Brazil and Venezuela, both important countries of manioc production and consumption, the roots are often processed by a mechanized version of an ancient hand process. The roots have traditionally first been shredded and pounded to ensure all of the cells are ruptured and all of the enzymes and glycosides have been released. The soggy mass is drained and the resultant wet fibrous residue allowed to sit overnight. Chemical studies have shown that right after shredding and pounding, there are still large amount of toxic compounds in the mass.

Since some of the volatile hydrogen cyanide has had time to escape, the concentration are lessened after standing. The wet, day-old pulp can be used directly to make a kind of flatbread by spreading it on a large circular griddle about a meter in diameter. The circular manioc pancakes that are produced do not contain cyanide because the cooking drives off the last of the gas. These flat manioc breads, which taste something like shredded wheat, can be dried and powdered. If the powder is toasted, it is known as fafaro. For many natives of the Brazilian tropical lowlands, Fafaro constitutes the only food eaten during an entire day. The powder can taste like a cross between plain corn meal and flour to those people who are unaccustomed to it. Yet large numbers of people eat it simply by the handful. Where factory processing is not available, various other primitive methods are still employed. The roots are often ground by hand, and the wet pulp poured into basketry tubes 2 or more meters tall.

The baskets, which are hung from the top and closed at the bottom, are constructed in such a way that they act like Chinese finger puzzles. The constriction of the tubes caused by the weight of manioc, squeeze out the moisture from the tubes. This moisture runs down the baskets into vessels below. A modest amount of starch is present in the juice squeezed from the grated manioc mass. Although the juice is discarded, the starch can be recovered by letting it settle out of the solution and decanting that liquid. The starch cake left behind can be dried and eaten raw, baked into a breadlike product, or used commercially like corn or potato starch.

Purified manioc starch is a better sizing material than either of these two other starches, but it is not produced on a large enough scale to challenge them commercially. If the starch is heated on a hot metal plate or tossed in a hot metal drum, it becomes gelatinized into pellets known as tapioca. Tapioca is a common ingredient in fruit pies and puddings because it is a good thickening agent. The water solution drained from grated or pounded manioc can also be fermented into a beer that is perfectly sage to drink because the fermenting yeasts break down any traces of toxic compounds left in the liquid.

Sweet Potatoes

The thought that humans crossed the Pacific from the New to the Old World long before Columbus discovered America has intrigued people for years. Thor Hyerdahl, in his attempt to prove the possibility of such a journey sailed westward from the coast of Peru in 1947. He built his balsa raft, fashioned after the rafts of Oru Indians on Lack Titicaca in Bolivia. His success heightened the controversy as to whether prehistoric exchangers had occurred between South America and the islands of Oceania. The cultivation of sweet potatoes (*Ipomoea batatas*, Convolvutceae) by native peoples of Malaysia and Polynesia as well as by those of South America was one of the evidences in support to the hypothesis. There is little doubt that the species is native to South America. Fossilized sweet potatoes from the Andes have been dated as being 8,000 and 10,000 years old.

Archaeological evidence has shown that they were cultivatred in South America by at least 2500 B.C. The unsolved question is when and how the sweet potato reached the Old World. From 1200 A.D., sweet potato forms part of an agricultural complex in Polynesia, although fossil and archaeological remains from the Pacific Islands are not available. Ancient pits believed to have been used to store sweet potatoes have also been found in New Zealand and roots figure prominently in the mythology of primitive New Zealand peoples in much the same was as true yams (*Dioscoreaceae*) do in Melanesia and New Guinea. Was the sweet potato really carried across the Pacific by humans before Europeans began to traffic the oceans? The answer currently seems to be no. Unless fossil evidence of sweet potatoes establish their presence in Oceania prior to seventeenth century, this conclusion will undoubtedly stand.

Even if such evidence is found, it will not provide the answer as to how the roots crossed the ocean. Today, the sweet potato is cultivated throughout the world. In temperate areas it is grown as an annual, and in the tropics it persists as a perennial. The parts themselves are trailing vines with morning glory flowers. The fact that sweet potatoes cannot be propagated by planting pieces of the root in the ground, provided the evidence that these are true roots, although in the eighteenth century they were confused. with white potatoes due to their resemblanc to tubers. A whole sweet potato would have to be planted to get a new plant. Pieces of aerial stems are thus made use in plantings. Due to deposition of lignin in the cell walls regular roots become tough and hard as they mature. Lignification in the storage roots appears to be prevented by the same hormones that initiate swelling process. Additional cambium layers are produced by storage roots during swelling, that make up the cortical cell in which starch is stored.

Unlike manioc, sweet potatoes contain a modest of protein (about 2 percent by weight), some sugar, and almost as much carotene (tabulated as vitamin A) as carrots. Being much less resistant to rot and insect attack than manioc, sweet potatoes are less extensively grown than manioc.

Improvements resulting from breeding programs have been slow because of the difficulty of programs have been slow because of the difficulty of producing glowers under temperate summer conditions. In the United States, sweet potatoes are eaten like white potatoes or "candied" (cooked with sugar). In the southern and western portions of the country, the roots are often called yams, which causes some confusion with the true yams. In the south, where varieties of sweet potatoes with moist yellow orange flesh are usually eaten, the northern part prefers varieties with drier, lighter yellow coloured flesh. In other parts of the world such as Japan over half of the annual crop is used for starch, wine and alcohol. A relatively large percentage is also used for animal food. Ironically, China is now the world's leading producer of sweet potatoes, the species being native to America and the tropical regions.

SWEETS FROM STEMS AND ROOTS

When we eat candy or pour sugar into our coffee, we seldom think about the fact that sugar comes from the stem or root of a plant. Historically, most sweeteners used by humans (other than honey) have been syrups or crystallized sugar obtained by boiling down the sap or juice expressed from the stem of various plant. These include most table sugar and sweet syrups from cane sorghum, date palms and trees such as the sugar maple. Since sugar is by far the most important sweetener in the modern world, we will discuss it in some detail. What we commonly call sugar is sucrose, a disaccharide of glucose and fructose. Sucrose tastes sweeter to humans than either of its constituent monosaccharides.

It is for this reason that sugar tastes sweeter than corn syrup, which is a mixture of glucose and fructose. Most sugar is obtained from cane, but in some areas, sugar beets are used for its production. The final product, refined (white) sugar, is identical regardless of whether cane or sugar beets are use. Consequently processing of both is discussed together.

Sugarcane

Sugaracane is a perennial grass (*Saccharum officinarum*, Poaceae) believed to have been selected form wild species of *saccharum* by primitive peoples of the Far East, probably New Guinea. The process of producing sugar from cane has been known in India since 3000 B.C. Alexander the Great brought back to Europe stories of a honey from reeds, but since the Far East was the only area of the Old World where cane grew well, very little arrived in Europe. Honey was the primary sweetener in Europe until almost 1500. Sugar was initially so scarce in Europe that it was used primarily to make medicines more palatable (a practice still used for children' medicines).

The discovery of the New World provided Europeans with land that could grow tropical and subtropical crops such as sugar cane. Columbus brought cane to the New world on his second voyage (1493) and successfully established it cultivation in the West Indies. Sugar rapidly became an important American export, and its production was intimately linked with the history of the New World. The famous American sugar triangle of the eighteenth century involved New England, Africa, and the West Indies.

Raw sugar of molasses produced in the West Indies shipped to Connecticut, where it was used to make rum. The rum was sent to Africa to buy slaves, who were brought to the West Indies to provide labour for the cane fields. Payment for the slaves was in cane products that again went to New England. In 1764, the British Parliament imposed a tax on sugar with the passage of the sugar taxas. As a result, people in Connecticut started to smuggle sugar into the colonies. When British customs vessels began to patrol the waters, one the *Gaspee*, was burned and sunk. This overt act of aggression, resulting form the sugar Act, predated the Boston Tea Party (1773) as the initial piece of violence that finally led to the American Revolution. One of the reasons for the successful establishment of cane, once suitable areas were found, is the method by which it is grown. Virtually all sugar cane is propagated vegetatively by means of *setts*. A part of the stem which includes lateral buds and circle of cell which give rise to a number of adventitious roots. The initial adventitious roots absorb nutrients until new roots are formed. Once established, the cane puts out tillers which give rise to new plants. This method of growth allows the harvesting of multiple crops per year. After several years, yield per plant decrease and new field is planted.

Sugar Beets

Beet greens and beet roots (*Beta vulgaris*, Chenopodiaceae) were eaten by the Roman, but it was not until the eighteenth century that the Germans discovered that the roots contained appreciable amounts of sugar. In 1747, Andreas Margraff developed the first process to extract the sugar, but a truly productive extraction process did not become available until 130 years later.

At the same time, artificial selection began to lead with higher and higher sugar contents. Napoleon I realized the value of a domestic source of sugar and ordered research to be conducted on the plant. Subsequent work so increased the sugar content that the French achievements began to draw the attention of other temperate countries. In the 1890s, sugar beet cultivation expanded rapidly in the United States and throughout Europe. The sugar beet represents one of the greatest triumphs of plant selection.

Under the encouragement of artificial selection, the sucrose content of sugar beets has risen from a lowly 2 percent on the eighteenth century to over 20 percent today. Although tough and basically inedible for humans, sugar beets now provide most of the sugar for many European countries. In the United States, they are important in some regions, but production is limited by the availability and capacity of local processing plants. Since the United States has a ready supply of cane sugar from Hawaii, Puerot Rico, and the Gulf Coastal States, there has not been any great pressure to expand sugar beet production. The predominant use of cane sugar in the United States is therefore, a matter of economics, and not because of any inherent superiority of cane over beet sugar.

Sugar Refining

Crystallized sugar is made by processing the juice of cut sugarcenes or harvested sugar beets. Cane is ready to harvest 22 to 24 months after planting the setts. The canes can simply be cut or the fields can be burned to remove the leaves and evaporate much of the water in the stems, there by reducing the time needed later to concentrate the expressed juice. Sugar beets are harvested in the fall of the year they are planted. When they are harvested, they are often "topped" (the top of the root and the rosette of leaves are severed from the root). The tops are used as animal food. Once canes reach the mills, they are fed through rollers that crush them and express the juice.

Sugar beets are washed, shredded, and crushed. From this point on the process of sugar production from the sweet juice of either crop is essential identical. The expressed juice is usually heated with calcium hydroxide and water and allowed to cool and settle. The upper layer of clarified juice is siphoned off and sent to an evaporater. The muddy lower of clarified juice is siphoned off and sent to an evaporater. Heat and vacuum evaporaters are used to thicken the syrup until it has begun to crystallize. This partially crystallized sugar syrup is called the *massecuite*. When crystallization has reached a maximum, the crystals of raw sugar are separated out by centrifugation. The brown liquid separated from the crystals is molesses. Most sugar-producing areas ship raw sugar, or use part of it to make rum.

Consequently the modern rum-producing countries are also major cane sugar producers. Raw cane sugar is crystalline, but it is brown and not free-flowing. The refining of the raw sugar into whiter table sugar is accomplished in refineries located in the sugar-importing countries. The refining of raw sugar involves several steps. First, the crystals are redissolved into a heavy syrup and centrifuged to remove some of the remaining unwanted chemicals. Next, a filtering agent is added to the syrup which clarifies it as it passes through the liquid. Another filtration, usually through charred bone, produces enough clarity to allow final crystallization and production of the various forms of white sugar.

Granulated sugar is made by crystallization of the syrup mass in large pans, followed by spinning the rough crystals in hot, revolving drums. Powdered sugar is made by finely grinding the crystals. Because it prevents clumping, corn starch is added to powdered sugar to produce confectioner's sugar. Sugar cubes are formed by slightly moistening the sugar and pressing it into forms. Despite the claims that raw sugar is "better" for you than refined sugar, the major ingredients used in the refining process, diatomaceous earth and bone char, are both innocuous substances. Most of the unhealthy effects of sugar are caused by the ingestion of too much sucrose. Raw sugar still contains many chemicals present in the cane stalks, none which have ever been shown to have any beneficial effects. Actually, what is often sold as "raw sugar" is refined sugar to which molasses has been readded.

Other Soures of Sweeteners

The making of sorghum "molasses" or corn syrup is similar to that of making sugar from cane except that the sugar is not crystallized from the syrupy juice. Sorghum (*Sorghum bicolour*) and corn (*Zea mays*) are both grasses like cane, but they are known primarily for their grains. Compared to sugarcane, both of these grasses are relatively little used as producers of sweeteners. In the Old World, primitive people of the Mediterranean region often collected the sap from palm trees (*Phoenix dactylifera*, Arecaceae) and condensed it by boiling off part of the water in order to produce a sweet syrup.

Likewise, early American settlers used the sap of native plants for syrup production. Importation of sugar was costly, and honeybees were not present, in North America until they were established by Europeans. Learning from the native Indians, the setters began to slash the bark of various trees, particularly the sugar maple (*Acer saccharum*, Aceraceae) and to collect the sap in the spring as it exuded from the cut. The sap was then boiled down until it reached the desired consistency. The amount of sap of temperate trees needed, however is enormous compared to the amount of juice form crushed cane because the sugar content of sap is very much lower (maple sap has 8 percent sugar) than of cane juice (about 22 percent). It takes approximately 40 gallons of sugar maple sap to produce a gallon of syrup.

3

Forest Products : Wood and Cork

Various products are obtained from forests useful for the making. In this chapter only two of the numerous forest products, wood and cork, will be discussed whereas the others are dealt with in other chapters.

Wood

Importance of Wood

After food and clothing wood is quite important for human being. Processed and unprocessed wood finds several uses in our day-to-day life. As a raw material for construction purposes it is highly versatile. It also gives several conversion products. It is cheap, light, stiff, strong, hard and tough material.

Structure of Wood

Botanically the term wood is applied to the secondary xylem produced by the cambium in the stems of gymnosperms and dicotyledonous angiosperms.

Soft wood and hard wood are the types, the wood can be classified into *nonporous* or *porous*. In Gymnosperms the wood usually consists of *tracheids* and is termed *soft wood*. The tracheids perform the function of both mechanical support as well as conduction. It is called *non-porous* as it lacks vessels. In Angiosperms, the wood is termed *hard wood* and it consists chiefly of two kind of elements—*wood fibres* or tracheids which give mechanical strength and *vessels*, which carry on conduction of

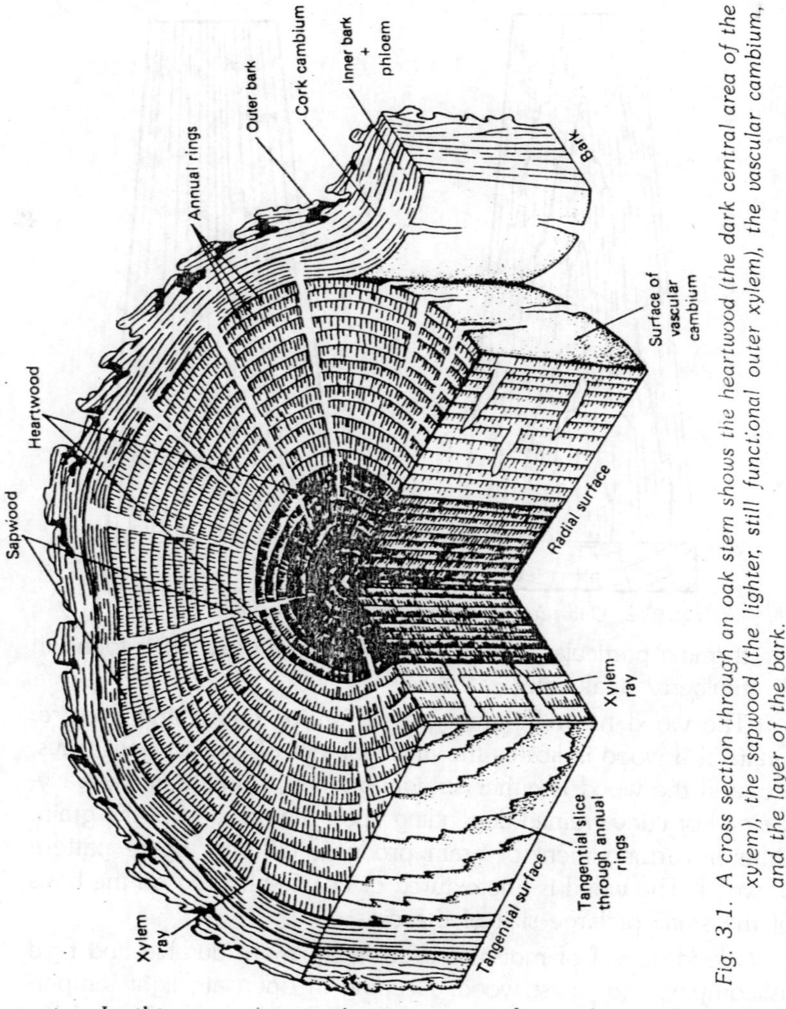

Fig. 3.1. A cross section through an oak stem shows the heartwood (the dark central area of the xylem), the sapwood (the lighter, still functional outer xylem), the vascular cambium, and the layer of the bark.

water. In this case due to the presence of vessels wood is called *parous*. Besides these elements the living *parenchyma cells* present in the wood perform the function of distribution and storage of carbohydrate food.

The porous wood is of two kinds :

(a) *Diffuse porous* in which all the vessels are of uniform size. This type of wood occurs in mango, shisham and sal.

(b) *Ring porous* in which the vessels are different in size in the spring wood (*early wood*) and autumn wood (*late wood*)

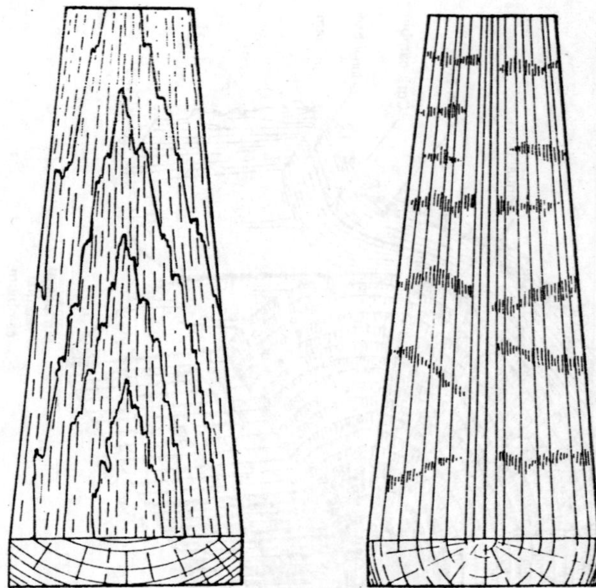

Fig. 3.2. Quartersawn (right) and plainsawn (left) boards.

forming particular annual rings. This type of wood is found in mulberry, teak, etc.

The wood has different types of *grain*, *figure* and *texture*. Grain of a wood indicates the arrangement of growth rings, rays, etc. and the wood is either straight-grained, spiral-grained, wavy-grained or curly-grained depending upon the position of the grain. Different arrangement of grain produces different figure pattern of wood. The wood is finetextured or coarse-textured on the basis of the small or large size of the cells.

Wood devoid of moisture is stronger, more durable and rigid as compared to moist wood. Keeping in open air, light temperature or in steam chambers are the different methods used for seasoning of wood to make it moist-free. Seasoned wood is particularly useful in the construction of furniture, door, show-case, etc.

Uses of Wood

There are a number of uses of wood. They can by broadly classified as follows :

Unprocessed Timber

1. Veneers, plywoods and laminated structures
2. Cooperage (barrels, tubs, etc.)

3. Fuelwood
4. Lumber and sawed timbers
   (a) Manufactures, including furniture
   (b) Construction, including maintenance and repair
   (c) Crating and shipping supplies
5. Ties and railroad maintenance
6. Posts, mine timbers, poles and pilings
7. Miscellaneous items

Wood Conversion Products

1. Plastics
2. Modiffied woods
3. Rayon
4. Pulp and paper
5. Distillation.
6. Hydrolysis, fermentation and other hemurgic products

A brief account of the above-mentioned uses of wood is given below :

UNPROCESSED TIMBER

Veneers, Plywoods and Laminated Structures

Veneers are thin sheets of wood cut from softened (boiled) logs by *rotary*, *slicing* or *sawing* process. An odd number of such sheets of veneer glued together at right angles to one another are called polywood. *Container veneer* is the cheapest quality of veneers. They are used for boxes, barrels, crates, baskets, etc. The *commercial veneers* are used for plywood, furniture, radio cabinet etc. *Face venners* are used for finest work. The plywood is used for panelling walls, flooring, partitions, doors, cabinets, shelves, furniture and a variety of other things.

Cooperage

Wooden staves are bound together for manufacturing a variety of containers. There are two kinds of cooperage—*slack cooperage* and *tight cooperage*. Slack cooperage is meant for holding and transporting solid substances like flour, fish, meat, vegetables, fruits tobacco, crockery, sugar, cement, etc. The tight cooperage includes barrels and kegs which are meant for transporting liquids such as deer, wine, oil and other liquids, paraffin is applied to the inner surface to prevent any leakage. Heavy cooperage includes huge tanks and vats.

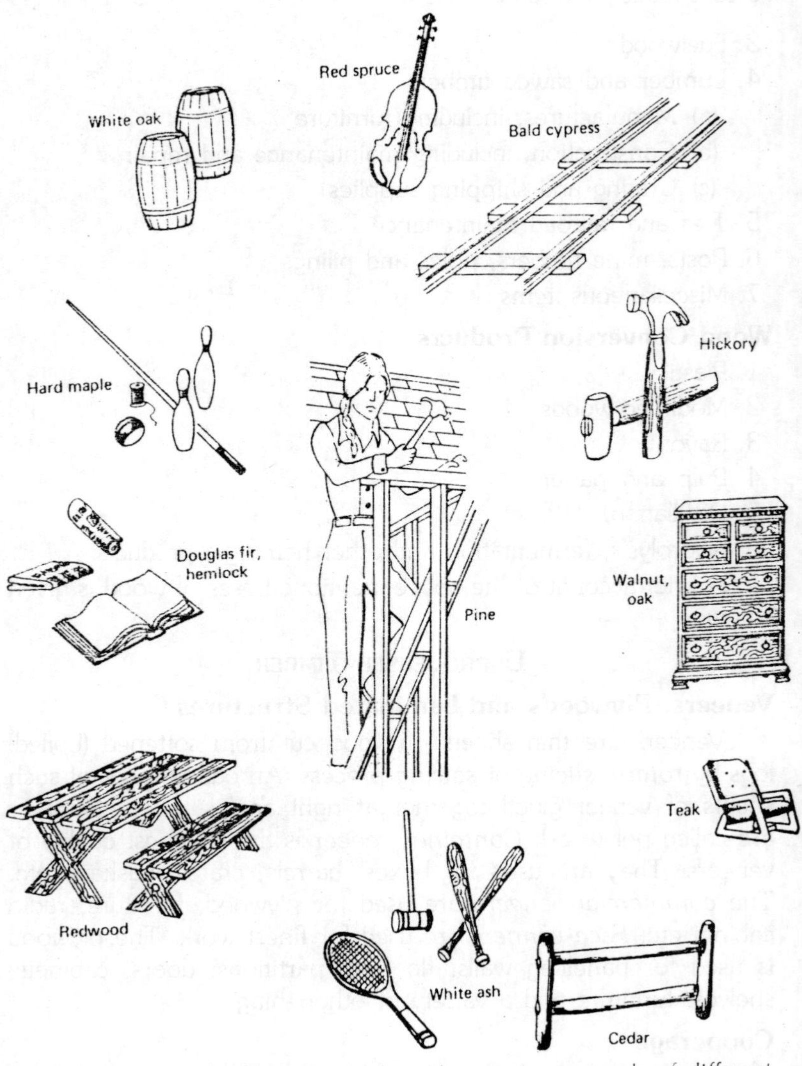

Fig. 3.3. Because of differences in their characteristics, woods of different species lend themselves to different purposes.

Fuel

Wood is extensively used for fuel particularly in rural areas. It is a very useful/and popular fuel since in dry condition it is 99 per cent combustible. Remains of plants fossilized into the form of *coal* are also a very widely used fuel. Destructive distillation of coal provide coal gas, benzol, naptha, ammonia, coal tar (source of dyes, antiseptics, etc.) and coke. *Lignite* is also used for fuel purposes. Accumulated deposits of vegetable matter in the bogs and swamps are slowly decomposed and carbonized and are termed *peat*. It is also used for fuel in some countries.

Petroleum, which is an important source of energy is a product of ancient plants and animals. Refined petrol is used in motor cars, scooters, and in industries. High quality gasoline obtained from fractional distillation of petroleum is used as aviation fuel. The other products are kerosene, jelly and paraffin.

Lumber and Sawed Timbers

Wooden logs cut for use are usually called 'lumber'. 'Timber' is the large sized lumber used for heavy construction or carpentry. An account of some timber plants is given at the end of this chapter. Lumber is used for a number of purposes.

Crating

Timber is very widely used for manufacturing crates, boxes, baskets, etc. for vegetables, canned goods, farm products, cosmetics, tea packets, machinery and instruments.

Manufactured products

A number of items are manufactured from wood. They include furniture, railway wagons, vehicles, boats, musical instruments, matches, toys, picture frames, agricultural implements and numerous other items.

Planning-mill products

Big lumber is recut into smaller sizes producing doors, blinds, window frames and interior finish like baseboards, columns, stairwork, posts, porch work, grills, etc. in the planning mills.

Structural purposes

Sawed timbers obtained from large-sized lumbers of preferably softwoods are used for the construction of residential and other buildings, sub-buildings, bridges and other types of heavy construction.

Ties and Railroad Maintenance

Uniformly sawed wooden slippers or ties treated with preservatives are extensively used below the rail tracks throughout the world. Despite the use of metallic ties these day, the wooden ties are still preferred most as they are cheap, elattic shock resistant, strong, durable and easy to transport. Such woods are also used in the manufacture of freight cars, trucks, and other similar vehicles.

Posts, Mine Timbers, Poles and Pilings

Wood is widely used for fencing posts around farms, along roads, railways, etc. Wood, mostly in the form of unsawed timbers, is used as props are supporting overhead burden in tunnels as *lagging* between the props and the walls to prevent their collapse and *ties* (foundations for rails): Poles used for telephone, telegraph, electricity and power-transmission lines are made up of non-conductive and non-corrosive cylindrical, durable sapwood.

Pilings, preferbly softwoods, are meant for bearing heavy top loads. They are used in building constructions, bridges, trestles and docks.

Miscellaneous Items

Excelsior or *wood fibre* are thin strands of wood used for packing and shipping glassware and other breakable articles and for stuffing mattresses and upholstery because of their lightness and resiliency. They are also odourless and free from dust and dirt. *Wood wool* used in the manufacture of rugs and matting, is a fine quality excelsior. *Sawdust* and *shavings* and used as fuel and as packing material. *Wood flour* produced by grounding wood wastes, sawdust and shavings is used in the manufacture of linoleum, plastics, explosives, veneer bonds and many other products. *Shingles* and *shakes* are thin sections of wood arranged in an overlapping manner on the roofs and sides of buildings to protect them from the weather.

Wood Conversion Products

Plastics and other Products

Cellulose nitrate formed by the action of concentrated nitric acid on cellulose gives a variety of useful products. They include *celluloid* and *plastics*, *guncotton*, *pyroxylin*, *collodion*, *varnish*, and *artificial fibres*. Wood, wood wasters, and wood flour are used in the manufacture of plastics. The *lignin* is believed to be responsible for the physical properties of plastics and other similar substances. The first plastic, celluloid, was made from cellulose nitrate and natural camphor. Today plastic products occupy a very important place in our daily life. A large variety of plastic articles are manufactured these days. Billiard balls, toys, handles, etc. are manufactured from celluloid and .is dyed to give a number of colourful objects.

Modified Woods

Wood is treated with resins and is then heated to setting temperatures and is finally compressed in press to give modified woods.The modified woods are dark coloured, heavy, hard and durable but are also very brittle. Panelling, airplane propellers, pulley and gear wheels, insulators, etc. refer to various applications of modified wood.

Rayon

Rayon, the yarn also called as artificial silk is the textile fibres derived from cellulose or chemical pulp and cotton linters. Cotton linters are short hairs removed after ginning in the cotton mills or yarn which is called rayon or 'artificial silk'. The cellulose is dissolved and the resulting liquid is then drawn into long strands or threads which are hardened to give the rayon fibres. Chemically rayon is, therefore, cellulose or an ester of cellulose. There are four kinds of rayon produced from four kinds of chemical processes each differing from the other in the solvent used for dissolving the cellulose. They are (i) *Viscose rayon* (ii) *Cellulose acetate rayon*, (iii) *Cuprammonium rayon*, and (iv) *Nitrocellulose rayon*. The first, third and fourth are pure cellulose while the second one is an easter of cellulose. Rayon yarn extensively used as a textile fibre and next to cotton most widely used in the manufacture of cloth. It can be more easily dyed than the natural fibres. Various household and industrial textiles are manufactured from Rayon.

Paper

Paper has a very interesting history. The word paper is derived from 'papyrus' a vegetative parchment. It was prepared using pith of a plant *Cyperus papurus* by the ancient Egyptians in 2400 B.C. The Romans are also known to have used the Egyptian papyrus for writing purposes. The first makers of paper were, however, the Chinese. The paper industry spread to India, Persia, and Arabia from China. Later it spread to Spain and other European countries. In 1690 a paper mill was established in U. S. A. The Chinese made paper from pulp prepared by boiling old rags, silk waste, hemp, rice leaves and baek of mulberry trees. The invention of a machine by French worker *Louis Robert* in 1799 for manufacturing paper and the discovery by *Frederic*

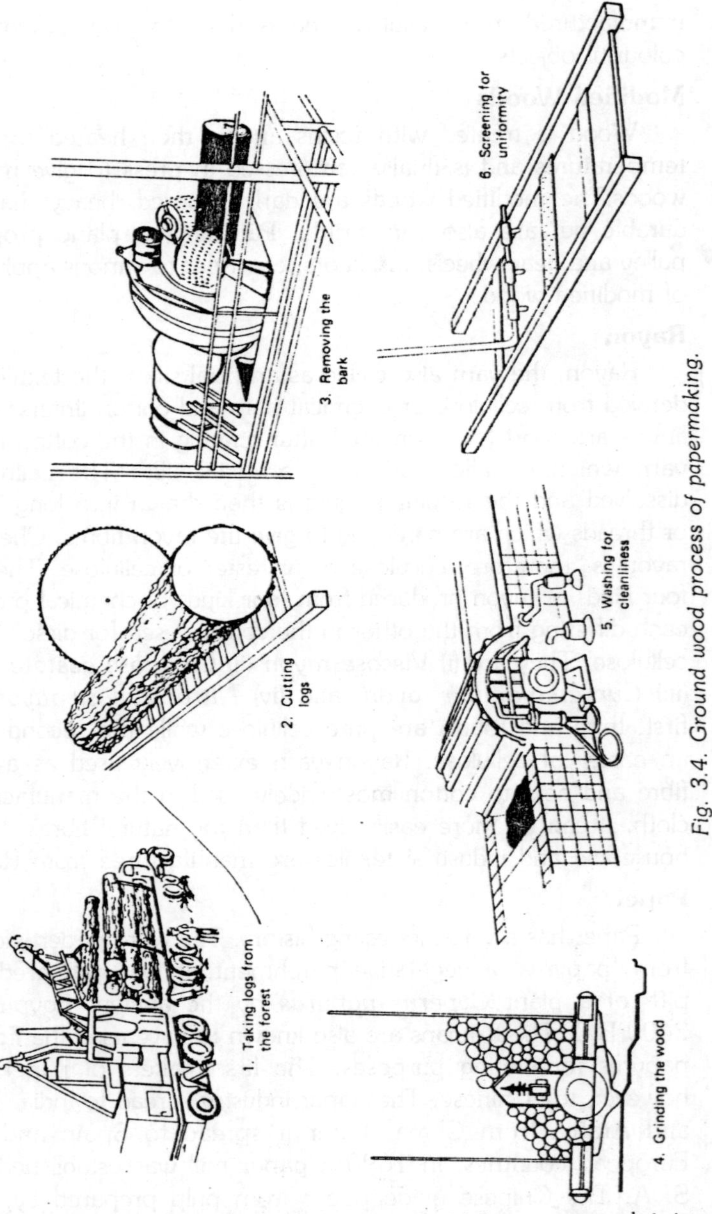

Fig. 3.4. Ground wood process of papermaking.

Keller in 1845 that ordinary wood could by converted into a paper pulp contributed to the establichement of the modern paper industry.

Raw materials

The chief raw materials of paper are natural fibrous materials like wood fibres, cotton and linen. Cellulose in the pure form gives the best quality paper. Wood pulp, composed of long and strong cellulose fibres is the most wanted raw materials for the paper industry. Spruces (*Picea* spp.), pines (*Pinus* spp.), aspens (*Populus* spp.), and the balsam fir (*Abies balsamea*) are important wood pulp-yielding plants. Some hardwoods like beech (*Fagus grandifolia*), sugar maple (*Acer saccharum*) and birch (*Betula lutea*) also furnish provide wood pulp.

Before the establishment of wood pulp as chief raw material cotton and linen rags were the only source of paper and still continue to be used for producing fine papers. Cotton has a very high percentage of cellulose (91%) and linen (flax fibre) contains 82% pectocellulose. Other raw materials of minor importance are textile fibres of jute and hemp, wastes of the textile industry, the esparto grass (*Stipa tenacissima*), paper mulberry, papyrus (*Cyperus papyrus*), bamboo (*Bembusa* spp.), baobab (*Adansonia digitata*) *Dephne cannabina* and *Eulapiopsis binata* (Baib-grass) and a number of other plants and fibrous materials. Low-grade paper, strawboard, cardboard and pasteboard etc. are produced from the fibrous stems of heat, rye, barley, rice, oats and other grasses.

Manufacture of wood pulp

The wood pulp industry is very large. The conversion of wood into pulp is brought about by mechanical or chemical processes. The mechanical process employed in the manufacture of poor grade papers which are not very good in strength, durability and printing. It turns yellow in colour as it contain resins, lignin etc. The newsprint is manufactured by this process. The chemical processes remove lignin and other undesirable things.

There are three kinds of chemical proceses (i) *soda process*, (ii) *sulphite process* and *sulphate process*. Sods process is the oldest one and is practised for the reduction of hardwoods and pine by alkaline cooking. The sulphite process involves cooking of wooden chips with steam in a solution of acid calcium sulphite. It turns softwoods of low resin content into a very stron pulp. 50 percent of the total pulp is formed in this way. The pulp gives high quality paper. The sulphate process is now widely used for

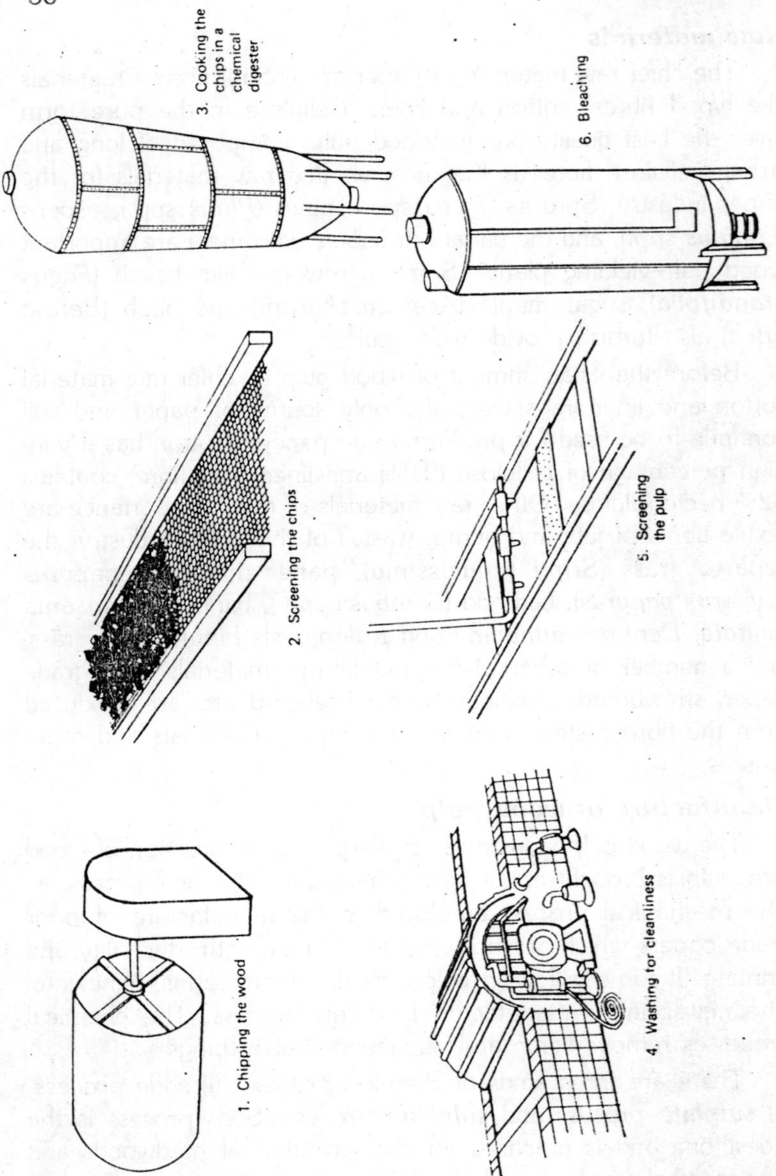

Fig. 3.5. Chemical process papermaking. (Continue...)

producing pulp from coniferous woods with a high resin content which is used in manufacturing a strong brown wrapping (kraft) and fibreboard containers.

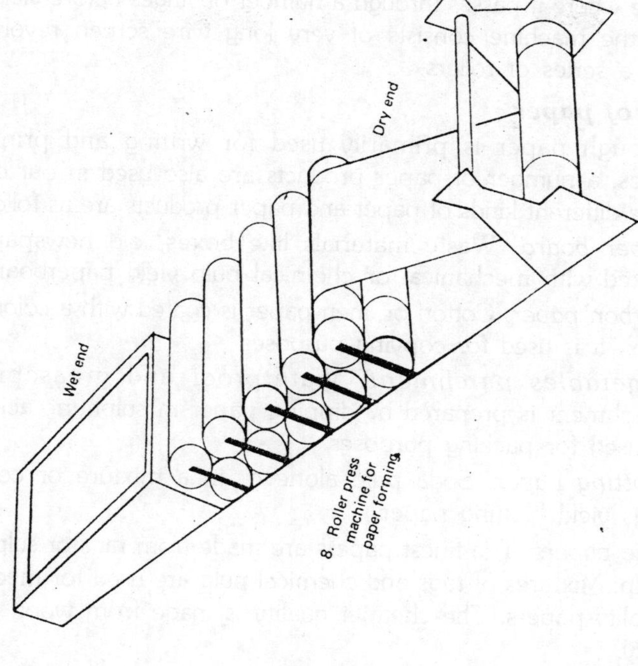

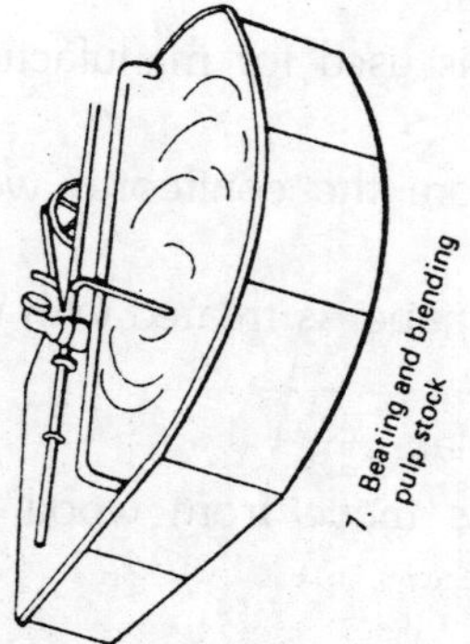

Fig. 3.5. Chemical process papermaking.

Manufacture of paper from pulp

The pulp is first bleached and washed. The pulp is then subjected to beating in a machine called beater or 'Hollander' to separate the fibres. Sizing to make the paper water and ink resistant and dye, if any, is added during this process to colour the paper. The pulp is then transferred to the papermaking machine where it passes through a number of stages before yielding paper. the machine consists of very long wire screen revolving around a series of rollers.

Kinds of paper

Though paper is primarily used for writing and printing purposes, a number of paper products are also used in our daily life. The different kinds of paper and paper products are as follows:

1. *Paper board*. Waste materials like boxes, old newspapers mixed with mechanical or chemical pulp yield paperboard.
2. *Carbon paper*. Cotton or linen paper is coated with a coloured wax. It is used for copying purposes.
3. *Vegetables parchment*. Waterproof and greaseproof parchment is prepared by dipping paper in sulphuric acid. It is used for packing purposes.
4. *Blotting paper*. Soda pulp alone or in a mixture or cotton rags yield blotting paper.
5. *Fine papers*. The finest papers are made from rags or sulphite pulp. Mixtures of rags and chemical pulp are used for medium quality papers. The cheaper quality is made from wood pulp only.
6. *Papreg*. The paper is mixed with plastics. It is moulded into a very strong but light weight material.
7. *Facial tissue paper*. Sulphur pulp is used for manufacturing this paper.
8. *Wrapping paper*. It is obtained from the conifeorus woods and some textile fibres.
9. *Roofing* and *building papers*. The paper is treated with wax, tar, oils, resins or other substances.
10. *Cigarette paper*. It is made from flax.
11. *Newsprints*. As stated earlier it is made from wood pulp prepared by the mechanical process.

Distillation

Distillation of softwood to produce turpentine and rosin. Hardwood is distilled to produce charcoal, acetic acid and wood alcohol (methanol). Charcoal is used as a fuel in many countries.

Cork

Cork is obatined from the bark of trees, particularly from cork-oak (*Quercus suber*). The tree grows inthe Mediterranean

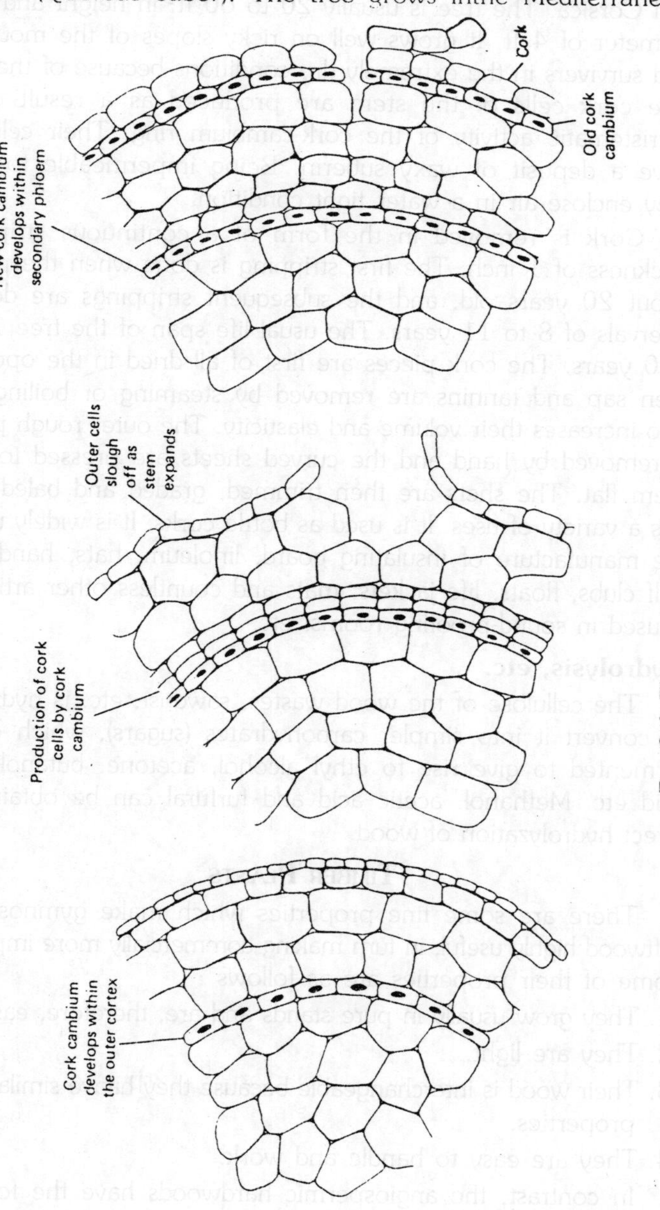

Fig. 3.6. The production of cork.

region, which is its native place. The cork-oak growing countries are Spain, Portugal southern France, italy, Algeria, Tunisia, Morocco and Corsica. The tree is usually 20 to 60 ft. in height and has a diameter of 4 ft. It grows well on ricky slopes of the mountains and survivers in the extremely dry conditions because of the cork. The cork cells of the stem are produced as a result of the meristematic activity of the cork-cambium ring. Their cell walls have a deposit of waxy suberin. Being impermeable to water they enclose air in a water tight condition.

Cork is removed in the form of a continuous strip of a thickness of 1 inch. The first stripping is done when the plant is about 20 years old, and the subsequent strippings are done at intervals of 8 to 11 years. The usual life span of the tree 100 to 500 years. The cork pieces are first of all dried in the open and then sap and tannins are removed by steaming or boiling. This also increases their volume and elasticity. The outer rough portion is removed by hand and the curved sheets are pressed to make them flat. The shets are then trimmed, graded and baled. Cork has a variety of uses. It is used as bottle corks. It is widely used in the manufacture of insulating board, linoleum, hats, handles for golf clubs, floats, life jackets, mats and countless other articles. It is used in sound-proofing rooms, etc.

Hydrolysis, etc.

The cellulose of the wood wastes, sawdust, etc. is hydrolysed to convert it into simpler carbohydrates (sugars), which can be fermented to give rise to ethyl alcohol, acetone, butanol, lactic acid etc. Methanol, acetic acid and furfural can be obtained by direct hydrolyzation of wood.

TIMBER PLANTS

There are some fine properties which make gymnospermic softwood highly useful, in turn making commercially more important. Some of their properties are as follows :

1. They grow usually in pure stands and are, therefore, easily cut.
2. They are light.
3. Their wood is interchangeable because they haave similar wood properties.
4. They are easy to handle and work.

In contrast, the angiospermic hardwoods have the following disadvantages:

1. There is great diversity in the qualities of wood and, therfore, this wood is not interchangeable.
2. They are heavy.
3. They hardly grow in pure stands. Only few plants grow in a highly mixed type of forest.
4. They are difficult to handle, transport and to work.

Softwood is usually sold as sawnwood for construction; roundwood for packaging purposes, for mining timber and fencing; veneerwood for plywood and pulpwood for paper and board industries. The hardwood has also similar uses. Decorative veneer work and furniture are the main used of high quality hardwood like that of teak. The important softwoods are *Pinus* spp. *Picea* spp. *Larix* spp., *Pseudotsuga* spp., *Cryptomeria japonica* and *Araucaria* spp., The hardwoods include *Betula* spp., *Eucalyptus* spp., *Fagus* sp. *Populus* spp. and *Tectona grandis*.

Sal (Shorea robusta)

Distribution. The sal tree is second only to teak in commercial importance in the India sub-continent In India there are two distinct sal belts which are separated by the Gangetic plant. The northern sud-Himalayan belt extends from Kangra valley in the Punjab to Assam. The second or central India velt extends from the Panchmarhi hills in the north-west through Rewash, Chota Nagpur, Midnapore to Godavari riven in the south. It belongs to the family *Dipterocarpaceae*.

Morphology. It is a large tree and is branched. It bears leaves which are glabrous, long and broad-ovate. The plant bears yellowish flowers in racemes in large, branched panicles.

Ecological factors. The plant can have maximum height of 3,000 ft. A mean annual rainfall between 40 and 100 inches and a temperature range between 55^0 and 88^{0}F are required for its growth. The plant grows well on loose and porous soils.

Quality and used of Sal wood

The salwood is not durable but the heartwood is very strong, heavy and tough. It has a cross-grained structure. It takes long time for sal timber to be seasoned and the dry and wet condition led to spliting and warping of sal timber. It is a very durable wood but not as much as the teak wood. The various uses of the sal timber are as follows :

1. The stem excludes an aromatic oleo-resinous gum 'sal dammar' which is used for caulking voats, in shoe polishes, carbon papers, typewriter ribbons, paints, varnishes, etc. It is also burnt as an incense. Sal dammer yields 'chua' oil on distillation. This oil is used for preparing 'attars' and incenses. It has some medicinal properties as well.
2. The bark of the tree is used as a tanning material.
3. The timber is widely used for pit-props, bridge construction, railway sleepers piles, boat building, masts, oars, poles, dyeing vats, carriages, cart wheels, ploughs, rice pounders, etc.
4. 'Sal butter' is obtained from the seeds. It is used as an illuminant, and as a substitute for cocoa butter in the manufacture of chocolates.
5. It yields charcoal which is used by blacksmiths.
6. Sal is also used for manufacturing inferior quality furniture, etc.

Shisham (Dalbergia sissoo)

Distribution. It is native to the syb-Himalayan tracts and the other Himalayan valleys, from Punjab to Assam. It also grows in Baluchistan, central India and Gujarat. It belongs to the family *Leguminosae*.

Morphology. It is large deciduous tree attaining a height of 60 ft. and more. Diameter of the trunk is 6—12 ft. It bears compound leaves, each having 3-5 leaflets. Yellowish white flowers are borne in short axillary panicles.

Ecological factors. The plant grows to a height of 3,000 ft. and sometimes 5,000 ft. It requires a very moist climate. It usually grows on sand or gravel along the banks of rivers.

Quality and uses of Shisham wood

Shisham wood is well known for strength and elasticity. It is widely used for the following purposes :

1. It is also used in the manufacture of various items like toys, articles for decoration purposes, combs, 'chappals', sports items, durms, etc.
2. It is used for interior decoration, cabinet work and various carpentary works.
3. It is very much used in building of boats, carts, carriages, agricultural implements, etc.

4. The heartwood yield an oil which cures skin and foot diseases of cattle.
5. It is used in the construction of buildings, for dors and windowframes.
6. The twigs and leaves are used as fodder for the cattle.
7. It is very good furniture-wood.

Teak (Vern. Sagwan) (Tectona grandis)

Distribution. The teak tree is native to central and southern India and southeast Asian islands. It grows well south to 25°N latitude in India, Burna, Malaysia, Indonesia, etc. In our country it is cultivated throughout Bengal, Assam, Sikkim and North-west India. It belongs to the family *Verbenaceae*.

Morphology. It is a large deciduous tree having four sided branches. The leaves are born in opposite manner. They are long, petiolate, elliptical, rough and glabrous. The plant bears small whtie flowers.

Ecological factors. Teak grows well up to a height of 3,000 ft. above which it has poor growth. It thrives well in mean annual temperature of 72° to 81°F. It requires a mean annual rainfall of 50" to 120". It grows on a great variety of soils. It grows on light and sandy soils, as well as on those which are binding and heavy. Perfect drainage and a dry subsoil are the necessities for teak plantations.

Quality and uses of Teak wood

Teak is the most durable of the commercial timbers of the tropics. In Burma, the tree is usually seasoned by a girdling process in which a deep circular cut is made through the brown-grey ba.k, the sapwood and the heartwood. The girdled tree dies after a few days. It is allowed to stand for one or two or even more years for seasoning. An outstanding quality of a seasoned teak timber is that it does not split, crack, shrink, warp, or alter its shape. Great durability of the teak wood is believed to be due to an aromatic oil in wood. It works easily and takes a fine polish. It does not corrode in contact with metal. The teak wood has a variety of valuable uses which are as follows :

1. Teak plants have other uses as well. The leaves are used as plates for thatching etc. A red dye is formed from leaves. The wood yields a pain-relieving substance. Flowers and the young fruits are believed to be diuretic. An oil is obtained

from the teak wood which is used as a varnish and as a substitue for linseed oil.

2. Teak wood is prized for high quality furniture, cabinet work, ornamental veneers, bowls, toys bodies of harmonium and piano and for numerous other things.
3. It is used for shipbuilding, boats, masts, etc.
4. It is widely used in construction of houses, bridges, piles, pit props in coal mines, railway wagons and carriages, railway sleepers, ploughs, and for cooperage work, etc.

Semal (Salmalia malabarica syn. Bombax malabaricum)

Distribution. It is native to India and Burma. It is usually cultivated. It grows up to an altitude of 6,000 ft. It is also called silkcotton tree. It belongs to the family Bombaceae.

Morphology. The plant is a very large tree. It attains a height of 150 ft. and a girth of 40 ft. The branches are horizontal and in whorls of 5 to 7. Conical prickles cover the young branches. The bark is smooth and silvery-grey. The leaves have 5 or 7 leaflets. Flowers are large, scarlet or white and are borne in axils of leaves. The fruit is an ovoid capsule which breaks into 5 segments. Numerous seeds with fine silky fibres are released after the dehiscence of the fruit.

Quality and uses of Semal wood

Having greysh white appearance, semal wood is coarse-grained and porous. The heartwood is not distinct. In dry condition the wood is not durable. It has several uses which are as follows :

1. The fine pieces of the semal wood called 'wood wool' is used as a packing material.
2. The stem secrete a gum (mocharas) which has medicinal properties.
3. It is used for planking, packing-cases, toys, scabbards, inferior quality pencils, cheap furniture, burush handles, box veneers, etc.
4. The seed yields an oil which is used as an illuminant and for making soaps, etc. The cake and the seeds are used as a cattle-feed.
5. It is a very good wood for match boxes and match splints.
6. The timber is used for piles which are immersed in water.
7. The silky fibres of the seeds are used for stuffing purposes.

8. Hollowed out trunks are usually used for making drums, water pipes, fishing-floats and rafts for floating and transporting heavier timbers.

Pine (Pinus spp.)

Pine timber (a soft wood) is an outstanding wood and accounts for almost one-half of the total lumber supply of the world. Out of several species from which wood is obtained, eight are important timber yielding plants. Broadly speaking the pines are classified into *soft* and *hard* types. The soft pine wood is straight-grained, soft, almost free of resin and is easy to work. The hard pine wood is resinous, heavy, hard, strong and durable. Some of the important pines are described below :

Pinus strobus (northern while pine of America)

It grows in north-eastern United States and parts of Canada. In England it is called Weymouth pine. The plant is a large tree and attains a height of 100 to 200 ft and a girth of 18 ft. The leaves are borne in clusters of five. Having white or slightly yellowish white appearance, the wood is very soft, light and easy to work. It is durable but has little strength. In northern America the timber is widely used for doors and windows, furniture, cabinet work, boxes, matches, etc. Other white pines of America are western white pine (*P. monticola*), the sugar pine (*P. Lambertiana*) and the lodgepole pine (*P. contorta*).

Pinus australis (southern yellow pine of Amierica)

It grows in the southern states of U.S.A. Heaviness, hardness, strength, stiffness and toughness of the wood are the best among softwoods. It is also very durable. The hard pine is widely used for beams, heavy constructions, bridges, ships, cars, railroad ties, boxes, etc.

***P. longifolia* (Vern. Chil, Chir)**

It is large tree with symmetrical branches. The leaves are in clusters of three. It grows in outer Himalayas and Siwaliks at altitudes of 1,500—7,500 ft. It is cultivated in the plains of north-west India. Chir is widely used for house building, oars, furniture, tea chests, boat building, general carpentry, sports items, match boxes, etc. The chir is believed to produce more turpentine and resin than the other coniferous trees of the north-west Himalayas. The crude turpentine and resin obtained by making incisions in the stem or by stripping off the bark are called '*Biroja*' or '*ganda firoza*', in north-west India, '*dhup*' in U. P., '*berja*' or '*binja*' in Garhwal, and '*khalija*' near Simla. '*Biroja*' has medicinal properties. The bark is used for tanning purposes. A dye used to dye yellowish shade to tassar silk and orange shade to wool is also obtained from bark.

Pinus excelsa (Vern. Kail)

It is also known as blue pine. It is a large evergreen tree. It has clusters of five needles. It is the best of our pines. It grows in the temperate Himalayas from Bhutan westwards at an altitude fo 6, 000 to 12,500 ft. Next to deodar it is quite durable. The wood is soft and easy to work. It is pinkish red and is moderately hard and highly resinous. 'Kail' is used for house buildings, shingles, water-channels, water troughs, wooden spades, packing cases, match boxes, match sticks body of violin, etc. It yields an excellent charcoal for iron-smelting. It yields turpentine, resin and tar. The bark is used for roofs of huts in the forests.

Deodar (Cedrus deodora)

Distribution. The deodar is limited in distribution. It is native to the mountains of Afghanistan and North Baluchistan, and to the northwest Himalayas. The plant is capable of growing to an altitude of 12,000 ft. It belongs to the *Coniferales*.

Morphology. The plant is a large evergreen tree. Attaining a maximum height of 150 ft, it is a tall tree and is very much branched. The smaller branches have drooping tips. The bark is dark coloured. Leaves are 1-1½ inch long, right and sharp. Cones are erect.

Quality and uses of Cedrus wood

The wood is yellowish—brown. It is oily and aromatic. Seasoning ofthe logs is accompanied by the decay of the sapwood. The heartwood is very durable and is almost imperishable in the climate of Kashmir and Punjab. The wood is light and keeps well in dry as well as wet conditions. Deodar like teak, shisham and sal is an important timber of our country and is put to various uses.

1. Destructive distillation of the wood yields 'deodar tar oil' which is antiseptic and is used for preserving skins.
2. The timber is widely used in the construction of barracks and other buildings, bridges, canals, railway sleepers, railway carriages, telegraph poles, etc.
3. It is also used for constructing storages vats for beer, packing cases, boxes, toys, furniture, carts, musical instruments, etc.

4

FIBRES

The fibre plants of the world are second only to the food plants in importance. Man has been dependent on these plants for his clothing and for a variety of other needs from times immemorial. The important fibre plants were cultivated during the ancient civilizations. The commercially important fibre plants are, however, not many. Botanically vegetable fibres are mostly made of sclerenchyma cells which are thick walled, slender and tapering with pits. Chemically the fibre cell walls are composed of cellulose, lignins, hemicelluloses, etc. Commercially the term fibre includes hairy or thread-structures borne by certain plant organs. Fibres are often classified according to their utilization: (i) Textile fibres, (ii) Brush fibres, (iii) Plaiting and rough weaving fibres, (iv) Natural fibres, and (v) Paper-making fibres.

Fibres excluding the wood fibre are of three types. The classification is based on their origin and structure:

(i) Soft, stem or bast fibres,

(ii) Hard, leaf or structural fibres,

(iii) Surface fibres.

SOFT OR BAST FIBRES

They are present in groups in the cortex of the dicotyledonous stems and are usually associated with the phloem (*bast*) or the pericycle. They are sclerenchymatous with deposits of lignin in the cell wall. The fibre or culsters of fibres are separated from the rest of the stem by retting in water. The examples of such fibres are jute, flax, ramie and hemp, etc.

JUTE (CORCHORUS SPP.)

Jute is a very valuable bast fibres and one of the most important commercial fibre crops of the word, particularly of India. They are obtained exclusively from two species of an Asiatic genus *Chorchorus* which belongs to *Tiliaceae*, a family closely related to *Malvaceae*. The genus *Corchorus* includes about 40 species. They are distributed throughout the tropical regions of Africa, America, Australia, China, Formosa, Bangla Desh, Sri Lanka, Nepal, India, Java, Japan and the Malayan Penisula etc. *Corchorus*

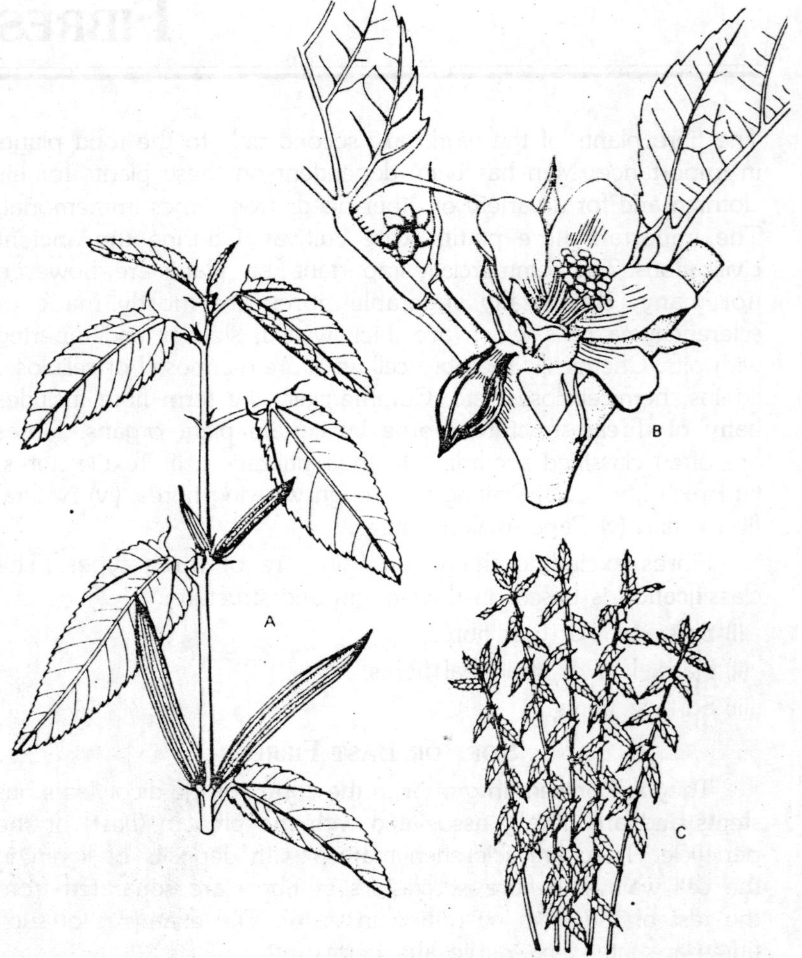

Fig. 4.1. Jute. A—Fruiting branch; B—Flower; C—Plant.

olitorius is believed to be of African origin and *C. capsularis* of Chinese or most likely of Indo-Burmese origin. The former is known to grow in Africa and West Asia for a very long time.

Both *olitorius* and *capsularis* have been domesticated in India in recent times.While *olitorius* has the finer fiber, *capsularis* is more adaptable particularly to flooded conditions. Jute is mostly confined to the Indian sub-continent. It was a monopoly crop for India till 1947 when most of the areas producing the finer quality of jute went to East Pakistan (Now Bangla Desh).

Raw jute and jute products are one of the main sources of foreign exchange earning for both the countries.West Bengal and East Bengal are leading place and produce nearly four-fifths of the world's total jute output. Its cultivation is being extended to some other parts of our country. There are 8 species in India out of which only two, *C. capsularis* and *C. olitorius* are of commercial importance. The other species are found wild in nature. *C. capsularis* is believed to the native of south China where it is found in the wild state. The plant might have entered India from there. This species accounts for nearly 75% of the total average under jute because of its ability to grow on both high and low lands and in water logged conditions. *C. olitorius*, on the other hand, can grow on highlands and cannot thrive under water logged conditions. The former species has about 50 and the latter 8 races.

The plant is a slender annual. It is quite tall and attains a height of 10 feet. Stem may be branched near the tip, but when planted close togehter they remain unbranched. It bears alternate, simple leaves. The plant bears small groups of two or three yellowish coloured flowers. *C. capsularis* bears small globular capsular fruits while *C. olitorius* bears long cylindrical capsules.

Ecological Factors

The jute crop requires a warm and humid climate. It's sowing season is March or a little later and the harvesting season is July-September. A temperature of 60°F to 100°F and high humidity is required for the cultivation of jute. The crop is often water logged but the young plants cannot withstand waterlogging. Jute is cultivated on the alluvial soils in the delta regions of rivers like Ganges and Brahmputra. The soils are of three kinds: light coloured.calcareous silts of fine texture, greyish soils rich in potash and old red alluvium.

Cultivation

The seed bed is prepared by ploughing and cross-ploughing the land about six times. *C. capsularis* can be sown at any time after the middle of February and with earlier sowings better crops are obtained *C. olitorius* is not normally sown before middle of April. Soil with fresh deposition do not need manuring. Other sols are usually manured with cowdung and wood ashes.Artificial manures are not used. The seeds are sown broadcast. During the first two months of the crop, a large number of plants have to thinned out gradually, so that the plants become 4-6 inches apart. Weedig is done with 'khurpi' two or three times.

Harvesting

Jute plants can be harvested when they are 3 to 4 months old. Harvesting of jute plant can be done at any time before they are dead ripe but harvesting is usually done at flowering stage. When the plants are in small pods it is considered as the ideal stage of harvest. At this stage fibers obtained are of good quality. If havested later the quality of the fibre is reduced because they become more lignified and woody to give coarser quality. The plants are cut close to the ground with a sickle. In high lands the harvested plants are cut and left in groups at different places in the field for 24 days when most of the leaves dry up. The plants are now tied in bundles of about 6-9 inches in diameter. At the time of bundling, the plants are shaken, when most of the leaves are shed on the ground. In many places the harvested plants are immediately tied into bundles which are then laid on the ground. In 2-4 days the leaves shed and the bundles are then taken for steeping in water.

Fibre Extraction

Retting. Retting is a process by which the fibres in the bark get loosened and separated from the woody stalk. The fibres as the pectin, gum and other mucilaginous substances are removed from it. All these process take place by the combined action of water and micro-organisms. Bundles are retting in water for 8-30 days. Retting can also be done by chemical process. Retting is completed when bark separates out from the wood.

Extraction. Extraction of fibre is done from the stalks of the retted jute by hand, jerks and pulls. The stems are often whipped on the surface of waer. Machines have also been designed and used in some countries for extracting fibres. The loosened i.e. the retted fibre are pulled apart from the sticks by hand. They are washed in water which is then squeezed out. The fibres are spread out in the sun on clean ground for 2 to 3 days to dry them. The dried fibrs are then tied together in small bundles.

Fibre Character

The jute fibre is developed in the outer portion of the bast (secondary phloem) of the stem. The fibre strand consists of a loose network of many smaller strands, each of which contains many elongated cells. The cells have pointed to tapering ends and are polygonal in cross-section. Cells i.e. the individual fibres vary from 2 to 5 millimeters in length and have wide lumen. The retted strands of jute are 5 to 10 feet in length. They are weaker than flax and hemp.They deteriorate rapidly in water. The quality of jute is reflected by its fineness, colour, lustre, length and strenght. The two species of *Corchorus* differ much in the quality of the fibre that they yield.

Better fibre is obtained from *C. capsularis*. The fibre of *C. olitorius* has a yellowish, reddish or greyish colour depending upon the nature of the retting water. The fibre of *C. capsularis* is whitish and the nature of the retting water. The fibre of *C. capsularis* is whitish and is called 'white jute' by the trade. The plant tissues contains tannin which combines with the iron of retting water and stains the fibre. There are a combines with the iron of retting water and stains the fibre. There are a number of well-defined varieites based on various characters : pigmentation, pod shape, petals and anther colour and yield and quality of fibre.

Uses of Jute

1. The jute stems, after extraction of fibre, are used for temporary fencing.
2. Jute seeds are rich in oil. 14.7% oil is extractable by petroleum ether. It can be used in the manufacture of soap and can also be converted into an edible oil. The oil from the seeds of *C. olitorius* can be separated into non-drying and drying portions the latter of which can be used in paints and varnishes.
3. They are also used in the manufacture of a kind of thick paper board.
4. Jute is a cheap and readily spun fibre. Jute fibre is the world's principal material for manufacturing coarse textile for sacs, bags and convases. The poorer fabrics are used in the making

of gunny bags while better ones are used for the manufacture of curtains, roofing fabrics and covers for cotton bales. The bags are used in the packing of sugar, potato, onions, pulses, coffee, cocoa, etc. It is used in the manufacture of twine, upholstery, rugs, carpets, etc.

5. The jute butts (pieces of lower ends of stalks and short fibres) are good source in the manufacturing of the paper.
6. The tender shoots are used as vegetable in Egypt, Sudan and other countries.

Flax (Vern. Alsi, Tisi) (*Linum usitatissimum*)

The flax fibres is obtained from the fibrous bundles of the stem of *L. usitatissiumum*, which belongs to family *Linaceae*. Linseed oil is also obtained from the flax fibre. The plant was well known to the primitive civilizations of Swiss Lake Dwellers, Hebrews, Egyptian, Greeks, and others dating back to well over 4,000 years. The Egyptians produced finest quality of linen and used it to cover their mummies. Linseed oil was used in embalming.

The early Greeks and Romans cultivated the plant for both the fiber and the seed. Ancient Indians used linseed oil in rituals. According to *Vavilov*, the small seeded types of linseed originated in South-Western Asia (Afghanistan). It spread northward to Europe and other parts of Asia and southward to India. The northern forms evolved into early races with long unbranched stems whereas the southern forms evolved into late races with lot of branching in the stem.

The fibre-flax are mostly produced in the north temperate regiosn of U.S.S.R., Poland, the Baltic States, Belgium, Holland and France. Some varieties are grown in warmer regions of India, USA, Canada, Argentina, Uruguay and USSR. Flax occupes the second or third place in usefulness amongst the fibres: cotton and jute ranking first and second. Flax is a much better fibre than cotton and gives rise to a finer fabric.

Cultivation

The fibre-yielding plant of flax is taller and attain height of 4 feet. Cool and moist climates of the temperates are required for its cultivation. In order to get long slender and unbranched plant they are spaced very closely. Harvesting can be done before the seeds are matured.

Fig. 4.2. Flax. A—Flowering branch; B—Seed capsule; C—Seed in section.

Extraction of Fibres

Retting. The plants are retted in two ways. The plants are usually exposed to dew for two to five weeks during which period the soft tissues decay. This process is called 'dew retting'. In cold water retting the plants are submerged in water of a pond, river, etc, for one or two weeks. Retting is quicker in water kept at controlled temperatures. Retting is completed in as short as 2 to 3 days and the quality of the fibre is also better. The fibres are loosened during retting process because the calcium pectate of the middle lamella dissolves.

Scutching. The retted plants are dried and the woody core is fragmented, crushed and beaten to separate it from fibres. The fibres (group of individual fibre cells) are 1 to 3 ft. in length. They appear yellowish in colour after extraction. The fibres are very fine–tough, flexible, strong, durable, long and lustrous. They are are stronger than cotton or wool. The cell wall contains 90 per cent cellulose. The fibres becomes stronger after absorbing water. They have qualities of strength, durability, gloss, water absorption and drying. The fibre is spun into quality yarn in the spinning mills.

Uses of Flax Fibre

1. Fragments of fibres separated during the scutching process are called 'scutchig tow' and are used in the manufacture of coarse fabrics and ropes.
2. The fibre is extensively used in the manufacture of convas, twine, carpets, cigarette paper, writing paper, firehouses, etc.
3. The fibres are chiefly used in the manufactures of linen cloth and thread. The threads are very strong and are, therefore, used for tailoring, shoemaking, manufacture of fishing lines and nets etc.

HEMP (VERN. BHANG) (*CANNABIS SATIVA*)

Origin and History

The true hemp (*C. sativa*) is one of the oldest of the cultivated plants. It is of Asiatic (central and western) origin. The plant is cultivated in China more than 4,500 years ago. According to *Schultes* (1970) hemp might have had originated in a diffuse manner in a huge area extending from Caspian and Himalayas to China and Siberia. The crop is believed to have spread to West Asia and Egypt during the period 1000–2000 B.C. and reached Europe during the Middle Ages (about 1500 B.C.).

Later it colonised the Mediterranean region. Hemp fabric has been found in Turkey (700–800 B.C.). It is now chiefly grown in U.S.S.R. and Europe (Itlay, Poland, etc.). India, Pakistan, Bangla Desh, Nepal, the countries of western Asia, China, Japan, African countries, southern states of U.S.A., Chile and Peru also grow hemp. Italian hemp is considered to be of the best quality. The varieties yielding best quality fibre were introduced in the western world from China. There are three kinds of varieites of hemp under cultivation: (i) Varieties grown in the Indian subcontinent and countries of western Asia for medicinal and narcotic properties. (ii) Varieties grown for the fibrous stems. (iii) Varieties grown for seeds whcih are edible and yield and drying oil. The yield '*marijuana*' or '*hashish*', '*ganja*', and '*bhang*' which have a hypnotic and hallucinating effect. They are derived from resinous secretions of the foliage and the inflorescence.

Botanical Description

Fig. 4.3. Hemp plant.

The hemp plant belongs to the family *Cannabinaceae*. It is an annual herb with a height of 3 to 16 feet. The slender stem is branched and bears large palmately compound leaves. The plant is dioecious. Male plants yield fibres of better quality.

Cultivation

Hemp grows well in a mild humid and warm climate. It requires an annual rainfall of 30 to 35 inches. Fertile loamy soil with rich amount of humus gives a very good yield of hemp. It is usually cultivated by sowing the seeds with the help of a drill. The plants for fibre purpose are closely spaced so that they remain unbranched.

The plants are cut just after they start flowering in order to get very strong and flexible fibres. The fibres are extracted in same way as the flax fibres extracted. All other process are done after "Dew retting" and scutching. The fibre occurs in the pericycle. It is white or dark brown. The fibres are similar to those of flax but are much longer–3 to 15 ft. in length. They are also stronger and more durable than the flax fibre. The flexibility and elasticity of the fibre is, however, less because of its lignification. The cell is composed of about 80 per cent of cellulose. the fibres is unaffected by water.

Uses of Hemp

1. The short fibres (*tow*) are used in the manufacture of caulk (*ooakum*), and as a packing for pumps, engines, etc.

2. Alkaloids obtained from hemp have medicinal and narcotic properties.
3. Hemp oil is drying oil obtained from the seeds of the plant.
4. The fine quality of the fibre is used in the manufacture of coarse fabrics which looks like a coarse linen cloth.
5. The fibre obtained from the stem is good source in the manufacturing of cordage (ropes and twines), sacking, carpets, tarpaulins, thread, binder twine, bags and webbing etc.

Other Soft Fibres

1. *Kenaf* (Verm. Patsan). The fibre is obtained from *Hibiscus cannabinus* grown in India, Fava, Nigeria, Natal and Egypt.
2. *Ramie* (China grass). The fibres are fine, long strong, durable and lustrous. The fibres is obtained from the bast of *Boehmerisa nivea* grown in China, Japan, Taiwan, Vietnam, Java and India.
3. *Sunn Hemp* (Vern. Sunn). The fibre is obtained from *Crotolaria juncea* which is an important Asiatic genus.

Surface Fibres

They are produced on the surface of stems, leaves, seeds, fruits, etc. This is the most important commercial textile fibre of the world. It is produced on the surface of seeds, whereas the stuffing material kapok is produced by the ovary or fruit wall on the inner side for the protection of developing seeds. Kapok is used in the manufacture of lifebelts. The husk of the coconut fruits yield the coir fibre.

Cotton (Vern. Kapas) (*Gossypium* spp.)

Origin and History

Cotton is the most important fibre in the world. The fibre is obtained from the seed hairs (*lint*) of a few of the several species (about 20) of *Gossypium*, which are native to both the Old and the New World. Cotton is known to have been under cultivation for weaving of its fibres as early as 4,500 to 5,000 years ago, both in Peru and in the Indus Valley of India. It finds references in the Rig Veda and in the laws of Manu, First of all the cotton fibre in the form of cloth was used by Hindus. *Gossypium* species are believed to have had several centers of origin.

Kinds of Cotton

It is very difficult to classify the numerous wild and cultivated species and varieties of cotton. The majority of species are found

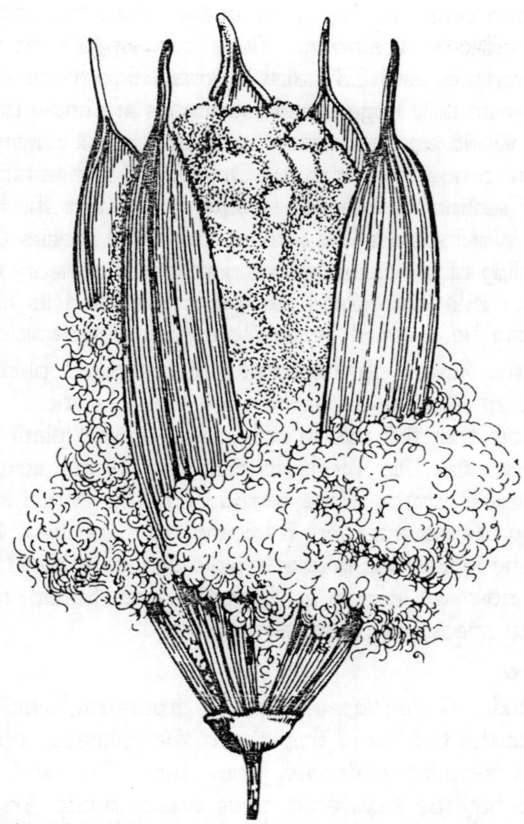

Fig. 4.4. Kapok pods split at maturity releasing their seeds covered with long silky hairs.

in the wild state and only four species are cultivated. The cultivated and wild species are either diploid or tetraploid with haploid chromosomes of 13 and 26 respectively. The tetraploids include only the New World cottons whereas all other species are diploid. There are seven section of diploid species depending on their place of origin and the particular genome they possess. The tetraploid species which are only three are kept in a separate section. The cultivated species belong to only two of the 8 sections, i.e. to section '*Herbacea*' and '*Hirsuta*' as shown below:

Herbacea:	Old World Cottons	(i) G. arboreum
	(Diploid)	(ii) G. herbaceum.
Hirsua:	Hew Word Cottons :	(i) G. hirsutum.
	(Tetraploid)	(ii) G. barbadense.

It is interesting to note that the New Word tetraploid sepecies of American origin do not grow in the wild state and, therefore, their ancestry in unknown. They contain 13 pairs of large chromosomomes and 23 pairs of small chromosomomes. Since no cotton with only large chromosomomes are know to come from the New World and no cotton with only small chromosomomes are known to have grown in the Old World. The tetraploid species of the VIII section are believed to have arisen from the hybridization of a New Word species (2x) with an Old Word species (2x) followed by a doubling of the chromosome number. This theory is supported by the fact that this hybrid can be formed and its chromosome number can be doubled artificially to give a tetraploid plant.

The half larger chromosomes of tetraploid plant are more like those of African *G. herbaceum* than Asiatic *G. arboreum*. This reflects that the cotton boll from diploid plant might have floated from the Old World to the New World across Atlantic. The tetraploid cotton were introduced into Africa in the 18th century by the slave traders returning from the New Word. They replaced the indigenous inferior diploid cottons in the 19th century. They are now well known as 'Egyptian' and 'Sudan' cottons. The Gossypium species are discussed below:

Herbacea

It includes *G. herbaceum* and *G. arboreum*, which occur wild throughout the Old Word tropics and the cultivated cottons of the Old Word have been derived from them. The wild species are perennial but the cultivated ones are annuals. The lint hairs produced by their seeds are smaller and are thus of poorer quality.

G. herbaceum is probably of African origin. It is extensively. cultivated in India, Iran, China, Japan and parts of Africa. It is a shrub with rigid stems and a height of 1-1½ meters. The bracteoles are rounded or broadly triangular, with margin divided into a number of triangular teeth. The cotton bolls are rounded. Their surface is smooth or is very shallowly dented. There are few or no oil glands. The bolls splits up into three or four loculi, each containing up to eleven seeds. The cotton is used for low quality fabrics, carpets and blankets. It can be very easily blended with wool.

G. arboreum (tree cotton) is either of Asiatic or African origin. It is grown in India, Persia and Africa from times immemorial. In the wild state it is a perennial, much branched shrub. It attains a

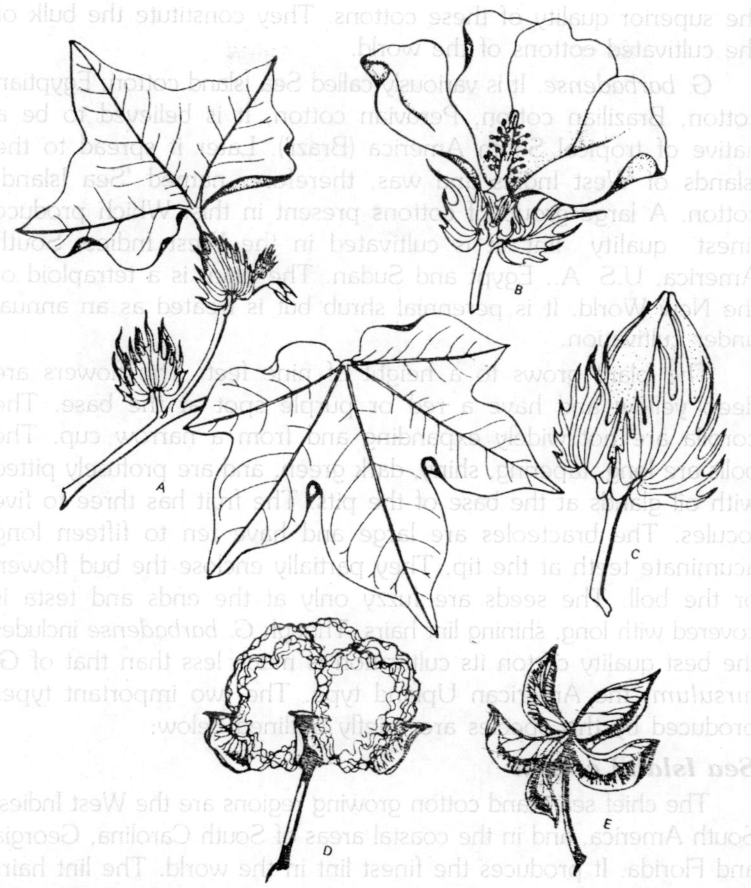

Fig. 4.5. Upland cotton; A—branch; B—flower; C—boll; D—open bcll; E—dried remnants of a ball with the seeds and fiber removed.

height of more than two meters and consists five to seven lobed leaves. The bracteoles are triangular, longer than broad, entire or with three or four coarse teeth. They invest the bud, flower and the boll very closely. Bolls are tapering and are profusely pitted with prominent oil glands present in the pits. They open widely when ripe and contain up to seventeen seeds per locule.

Hirsuta

This includes the tetraploid species of the New Word—*G. barbadense* and *G. hirsulum*. They originate in America and have spread to every cottons-growing country of the world because of the superior quality of these cottons. They constitute the bulk of the cultivated cottons of the world.

G. barbadense. It is variously called Sea Island cotton, Egyptian cotton, Brazilian cotton, Peruvian cotton, It is believed to be a native of tropical South America (Brazil). Later it spread to the islands of West Indies and was, therefore, named 'Sea Island' cotton. A large group of cottons present in this. Which produce finest quality lint. It is cultivated in the West Indies, South America, U.S. A., Egypt and Sudan. The plant is a tetraploid of the New World. It is perennial shrub but is treated as an annual under cultivation.

The plant grows to a height of nine feet. The flowers are deep yellow and have a red or purple spot at the base. The corolla are not widely expanding and from a narrow cup. The bolls are long, tapering, shiny, dark green, and are profusely pitted with oil glands at the base of the pits. The fruit has three to five locules. The bracteoles are large and have ten to fifteen long acuminate teeth at the tip. They partially enclose the bud flower, or the boll. The seeds are fuzzy only at the ends and testa is covered with long, shining lint hairs. Though *G. barbadense* includes the best quality cotton its cultivation is much less than that of *G. hirsulum*, the American Upland type. The two important types produced by this species are briefly outlined below:

Sea Island cotton

The chief sea island cotton growing regions are the West Indies, South America, and in the coastal areas of South Carolina, Georgia and Florida. It produces the finest lint in the world. The lint hairs are creamy white and are silky and lustrous. Staples are of more than 2 in. in length and are stronger and firmer than any other cotton. It is used for the finest textiles and yarns.

Egyptian cotton

It is chiefly cultivated under irrigation in the Nile Valley of Egypt and Sudan. It was introduced there from Central America. Egyptian cottons are of very mixed origin. The lint hairs are similar to those of Sea Island, but are shorter, measuring 1-3/8 to 1-3/4 in. in length They are brown coloured and are not so fine. This cotton is used for thread, underwear, hosiery, tire fabrics and fine dress goods.

G. hirsutum. It is believed to have originated in Guatemala of Southern Mexico of Central America. It is the so-called 'American

Upland' cotton. There are about 1200 varieties of *G. hirsutum*. They are widely grown throughout the great cotton belt of the southern United States, in India, East and South Africa and in Sudan under rain cultivation. The plant is a small shrub with very few vegetative branches. Leaves are occasionally palmatisect. Flowers are white or light yellow and unspotted. Petals are widely expanding. Bracteoles have numerous teeth like projections and closely invest the bolls. The ball appeared round and often large having pale green in colour. The surface is smooth. Oil glands are few and are sunk beneath the boll surface. The bolls are four or five-valved. The seeds are covered with short white *fuzz* and *lint* hairs of staple length of ½ in. to 1 in. (the 'short staple American Uplands') or of 1 in. to 1½ in. (the 'long staple American Uplands'). The lint is strong but not as fine or lustrous and strong as the lint of *G. barbadense*. It produces medium quality yarns and forms the bulk of the cotton trade of the world.

Botanical Description

Gossypium belongs to the family *Malvaceae*. They are herbaceous, shrubby or tree like plant. Under wild conditions they are perennial while the cultivated species have developed into annuals. The plants usually grow to a height of 2 to 6 feet. The plant has a long tap root, varying greatly in length depending on the type of soil and whetehr the crop is grown under conditions of rain cultivation or irrigation. The main stem is monopodial, erect and branched.

The leaves are *spirally* arranged on the vegetative branches of the stem and the leaves possesses two kinds of buds–auxillary and extra axillary. Usually only one of the two buds develop. The axillary buds of the of the lower nodes produce *vegetative* or monopodial branches whereas the extra-axillary buds of the upper nodes develop *fruiting* or sympodial branches. The vegetative branches may also bears fruiting branches. They are truely sympodial since a flower is borne at the tip and the growth is carried on by an axillary bud of a leaf. The fruiting branches are also different from the vegetative branches in the fact that they bear leaves in opposite manner in two rows. The leaves of cotton plant are variable in shape and size. The leaves are stipulate and have long petioles. The lamina is broad, palmately-lobed with three to nine lobes. The terminal buds of the sympodia of the fruiting branches gives rise to single flower. The flowers come to lie opposite the leaves. Normally, cotton is self-pollinated but cross-pollination may also take place. The fruits of cotton are called 'bolls'. The bolls are vaiable in shape and size. They are more or less egg-shaped and are pointed at the upper end. The bolls of *G. barbadense* are dark green, profusely pitted with large number of oil glands. The bolls of *G. hirsutum* are light green, smooth and have few oil glands.

The capsule of cotton fruits contains three to five locules which are commonly known as *locks*. When the locks separate the fibrous seeds are exposed. Each locule bears about nine seeds, and a boll may contain twentyfour to fifty seeds. The seeds are ovoid, more or less pointed, dark brown structures. The seed's coat produces two kinds of tubular outgrowths or hairs or hairs from the epidermal cells. The longer hairs which are thick and white or creamy are called the '*lint*' and the very short, white or coloured haris are termed the '*fuzz*'. While the lint or 'staple' is the cotton fibre of commerce, the fuzz is of little importance.

Fuzz is present all around the seed in most of the species, some have fuzz at only one end of the seed, and some do not have them at all. The fuzz hairs are strongly attached to the seed coat. The long tubular cell of the mature lint hair has has a very thick wall consisting of superimposed layers of spiral bands of cellulose. The hair is twisted at the points of "pits" which are thin spots in the wall. The number of twists in the fibre varies with different varieties. In the Indian cottons the average number of twist is about 150 per inch, whereas some of the best cottons may have as many as 300 twists per inch. The occurrence of twists in the cotton fibres enable them to be spun into yarn and is responsible for the coherence of the individual fibres.

New Varieties of Cotton. SH 131, SH 274, SH 2674, SH 175, 468-10, Varalakshmi, DCH-32

Quality of Cotton Fibres

The quality of cotton can be determined by several characters. The following characters are important in determining quality :

1. '*Counts*'. The quality of cotton is usually evaluated in terms of counts. A count is the number of hanks required to weigh a pound. A hank contains 840 yards of thread. Finer threads, therefore, have greater counts. Indian cottons are of about 22 counts whereas the best quality cottons may range from 80 to even counts.

2. *Fineness*. The fibres with very thick walls are 'coarse' and with thin walls ar silky and 'fine'.
3. *Ginning percentage*. This is the amount of lint obtained in relation to the amount of cotton fed into the gin. Obviously the variety of cotton which gives a higher ginning percentage is better.
4. *The length of the fibre*. The longer fibres are much better and preferred than shorter ones. The Sea Island cotton is, therefore, the best cotton grown inthe world.
5. '*Neppiness*'. During the process of ginning some fibres are entangled into knots due to failure to individual fibers to thicken during growth. This is a troublesome and undesirable character.
6. The *strength* of the fibre, that is, the actual strain that a fibre can withstand.
7. *Colour and lusture*. Cotton with rich bright creamy colour are better.
8. *The number of twists*. The larger the number of twists the stronger yarn obtained.
9. *Uniformity of the fibres*. The fibres of uniform length are easily spun into yarns and there is no undue wastage during manufacture.

Ecological Factors

Cotton is a tropical crop so it grows well in adequate but not excessive moisture during early stages of growth, a relatively dry period during flowering and no rainfall during harvesting. It requires a high and uniform temperature during the early stages of growth. There should be plenty of moisture in the soil which can be supplied by rain or irrigation. In the later stages of its growth the temperature should be lower preferably with cool nights. This checks vegetative growth and encourages fruiting. Boll opening and harvest requires dry weather. Sandy and moist soils are generally good for its cultivation. The river valleys are, therefore, very good regions for its growth. In India, cotton is cultivated in the black cotton soils of the Peninsular India, red soil of southern India and alluvial soils of the Indo-Gangetic plains.

Cultivation

The methods of cultivation are different in different soils and climatic conditions. Deep ploughing is generally not done. The land is usually harrowed 2 to 3 times to facilitate conservation of moisture in the soil. The seeds mixed with mud or cowdung are sown broadcast or in rows by drills. The distance between the lines depends upon the variety grown and the fertility of the soils and usually varies between 12. in and 36 in. While American cottons are grown pure, the local cottons are grown as a mixed crop. Crop rotation is beneficial as it adds fertility to the soil.

In northern India (Punjab, U.P. and Bihar) it usually alternates with wheat and jowar, in the black cotton soil tracts of peninsular India jowar precedes cotton. After germination the plants are thinned to 12 to 24 inches apart in the row. During thinning operation, 'roguing' or the removal of off-type plants is carried on. In black cotton sils when the plates are about 4 inches high a process called interculture is practised. The soil is cultivated with a bullock drawn daura or hoe to prevent damage from cracking. Weeding is done by hand in our country. In America weeding is done by tractor drawn cultivators, flame throwers, and chemical herbicides. The plans start flowering in 7-10 weeks time. The harvesting can be done either by hand or by machine and the seed cotton is picked from the open balls. In America mechanical 'strippers' or 'pickers' are used. A defoliant is applied shortly before harvest to facilitate mechanical stripping.

Preparation of Cotton Thread

This involves a number of operations. They are as follows :

1. *Carding*, combing and drawing. The short fibres are exracted, others are straightened and evenly distributed.
2. *Lapping*. It is a process of combining of three layers of cotton fibres into one.
3. *Twisting*. The fibres are twisted into threads.
4. *Ginning*. Ginning is done by a saw tooth gin or roller gin to remove the lint hairs from the seed and fuzz. The cotton seed, to which the fuzz is still attached but from which most of the lint have been removed is subjected to a further ginning process, and is then used either for seed or is sent to the oil press for oil extraction.
5. *Baling*. The lint hairs got from the ginning process are baled under pressure in bales for transporting to the mills.
6. *Picking*. In the textile mills a series of beaters and pickers fluffs the fibre and frees it of foreign particles and delives the cotton in a uniform layer.

Use of Cotton

1. The stalks of the cotton plant can be used as a fuel. It yields a fibre which can be used in paper making.
2. Cotton seed oil is one of the important semi-drying oils (25%). It is used for cooking purposes, and in soap making.
3. The *hulls* are a good stock-feed. They are used as fertilizer, and as a source of industrial solvents, explosives and many other things.
4. Cotton alone or mixed with others fibres are used in the manufacture of various types of textiles.
5. It is used in the manufacture of rubber-tire fabrics, in plastic reinforcing, carpeting and cordage.
6. The *linters* (fuzz) are used in stuffing pads, pillows, cushions, mattresses, etc. They also yields cellulose, absorbent cotton and low grade yarn. The yarn is used in the manufacture of twine, rope and carpets etc.
7. Absorbent cotton consisting of thoroughly cleaned fibres which are almost pure cellulose are basic raw materials for a number of cellulose industries.
8. The seed cake is a good sources of cattle or stock feed, and also acts as nitrogenous fertilizer as well as dyestuff.
9. Unspun cotton is commonly used for stuffing purposes.

Other Sarface Fibres

1. *Red silk cotton* (Vern. Simal). It is called Indian kapok. The fibre is obtained from Salmalia malabarica. It is used for stuffing purposes.
2. *White silk cotton*. It is a native of India is cultivated in tropics.
3. *Madar* (Vern Ak). The fibre is obtained from Calotropis gigantia.
4. *Kapok*. The pods of the kapod (Ceiba pentandra) bears floss. It is a very good silk cotton and is exclusively used as a filling fibre. It is used in stuffing pillows, cushions, furniture, mattresses, etc. The fibres are impervious to water and are five times more buoyant than cork and are, therefore, used in the manufacture of life-belts, life-buoys, portable pontoons, etc. They are widely used for insulating refrigerators and for making sound proof rooms. They have a number of other uses as well. The plant is grown in Java, the Philippines, Sri Lanka, India and Mexico. It grows wild in tropical America.
5. *Coconut* (Coir).

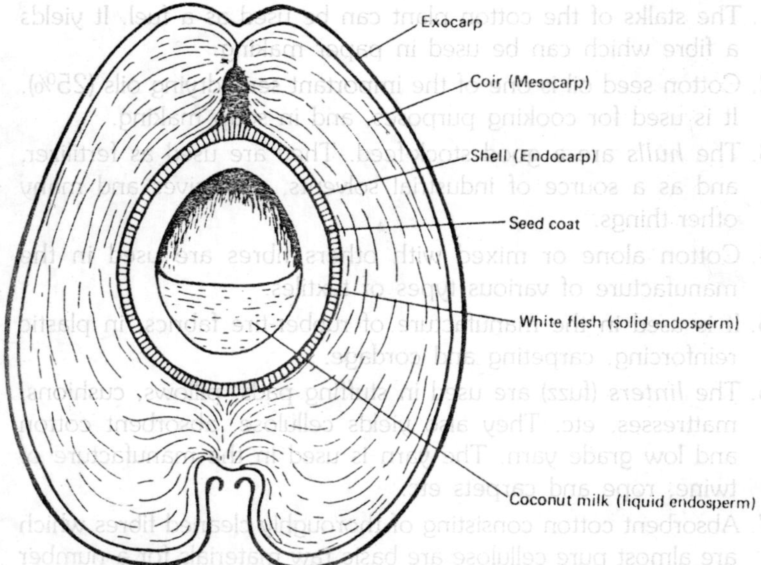

Fig. 4.6. Coconut fiber, or coir, comes from the fibrous fruit coat that surrounds the seed valued for its nutritions endosperm.

Hard or Leaf Fibes

They are found in the leaves of monocotyledonous plant's supportive and conductive strands. Hard fibres also consists of xylem and phloem so they are called *fibrovascular bundles*. The cells are very much lignified. The whole fibrovascular bundle serves a single unit. They are separated by mechanical scrapping. They form the *cordage* fibres of commerce and are used for the production of ship's cables, ropes and coarse sacking. Manila hemp, New Zealand hemp, sisal, Mauritius hemp, etc. belong to this category. A large number of plants produces hard fibres. They are as follows :

Maurititus Hemp

The fibre is obtained from the leaves of green aloe (*Furcraea gigantea*) growing in Mauritius, Madagascar, South Africa, India, Venezuela and Brazil.

Abaca or Manila Hemp

The fibre is obtained from the leaf stalks of several species of wild planntain or banana, particularly from *Musa textiles*. The plant is grown in India, Indonesia and the Philippines. This fibre is the best cordage material. It is very strong and elastic and is resistant to both fresh and salt water. It is used in the manufacture of marine cables, binder twine, bagging, wraping paper sacs, etc.

New Zealand Hemp

The fibre is obtained from the leaves of *Phormium tenax* growing in New Zealand.

Agave Fibres

Mexican Sisal (*Agave fourcroydes*) is a source of a agave fibre as its fleshy leaves produces the fibres, Sisal (*A. sisalana*), istle (*A. funkiana*), Manila maguey (*A. cantala*) etc.

5

FRUITS OF WARM REGIONS

Papayas, mangoes, sour sops, and custard apples bring to mind visions of humid tropical rain forests and thoughts of exotic flavours. Such fruits seem to have little in common with tomatoes, green peppers, squashes, and cucumbers. All the fruits are native to tropical and warm regions of the world, these fruits are unable to survive winters that have hard freezes. In agriculture we have circumvented the lack of winter hardiness in the cases of tomatoes, squashes, and peppers by planting them as annual crops that germinate, flowers, and produce harvestable fruit during the spring or summer season. These fruits are exceptions, however. Woody perennials are the source of most of the tropical fruits. These are unable to germinate and proeduce the fruits in a few months. Growth of such specie is therefore restricted to areas that do not experience freezes.

The tropics have many more kinds of edible fruits than temperate regions, yet temperate fruits dominate our markets and constitute an appreciable part of the commercially available fruits in tropical regions as well. The tropical fruits produced in very limited quantity because of their decaying nature. Due to which most are gathered from wild or cultivation is done on limited scale and used locally. A few exceptions include bananas, pineapples, and avocados, all of which have become readily available in the past century when methods of shipping therm without spoilage were developed. Luckily, the selection of tropical fruits offered in this country is expanding. We therefore begin this chapter with well-known fruits native to warm regions and then treat the more exotic fruits occasionally found on our grocer's shelves.

Citrus Fruits

Orange juice is an integral part of the "all-American breakfast." Yet, at the turn of the century, oranges were considered luxuries in northern markets and were prescribed as cold or consumption remedies by physicians. In the United States oranges are the largest perennial fruit trees. Oranges belongs to genus *Citrus*. All other relatives of oranges i.e. lemons, limes, citrons, grapefruits and tangerines all these also belongs to the same genus i.e. *Citrus*. All of these citrus species are trees with shiny, evergreen leaves. Most occur naturally in the region extending from north-central China southward to Australia and New Caledonia, but the cultivated citrus are believed to be derived from species native to southeastern Asia.

Placement of the domesticated members into biological species is difficult because hybridization and human selection for mutants has obscured natural boundaries. All citrus species share the same fruit type, a hesperidium, which is basically a berry with a leathery skin formed by the combined exocarp and mesocarp. The vegetative parts of the plants, the skin, the leaves of the plant are covered with pockets which is filled with aromatic oil. This aromatic oil filled pockets are the characteristics of the citrus family. The aroma given off when citrus peels are brushed, grated or twisted. The endocarp of citrus fruits is highly modified into fleshy hairs that fill the segments of the ovary and surround the seeds. These hairs, or juice sacs, are the parts of the oranges, grapefruits, and tangerines that we eat. They contain a watery solution of sugars and acids.

In some species, the acid concentrations decrease with age and produce fruits that we eat without addition of sugar. As earlier explained *Citrus* species are unable to tolerate real freezing temperatures and so these species are not natives of lowland and wet tropical regions. It requires good soil moisture, but abundant sun and reasonably dry air. In the past, most citrus fruits were grown from seed, but with the rise in popularity of certain cultivars, many of which are partially or completely sterile, bud grafting has become the most common method of propagation. Of the various species, bitter oranges, pummelos, and citrons are still frequently grown from seed. The most widely grown citrus fruit in

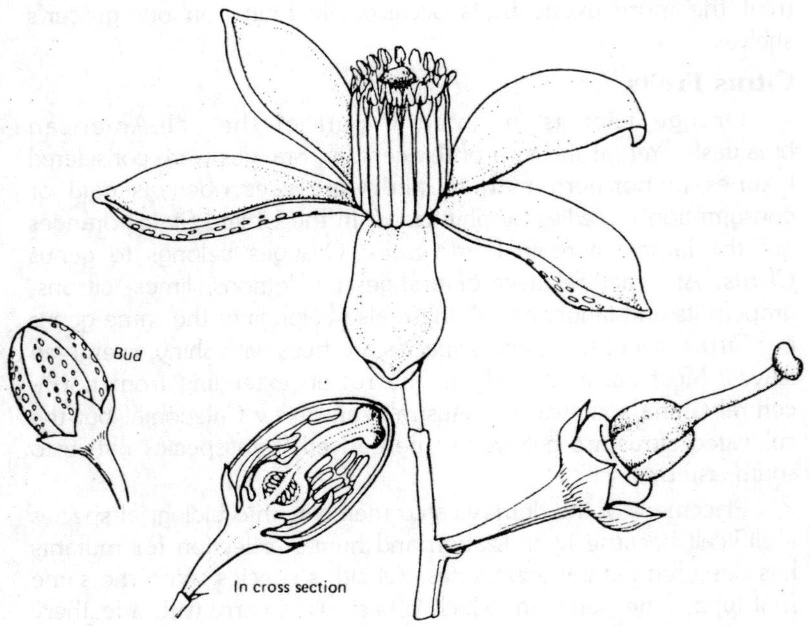

Fig. 5.1. Figure showing a branch bud and flower of a sweet orange.

the world is the sweet orange (*Citrus sinensis*), but with the wild ancestors gone, we can only speculate about its origin. Some authors believe the orange is derived from an unidentified or extinct Chines species but others assert that it resulted from selection of a hybrid between a tangerine and a pummelo.

Oranges were considered by some to be the "golden apples" of Greek mythology that the Goddess of Fertility gave to Hera when she married Zeus. Here planted the seeds in the garden of Hesperides, hence, the designation "hesperidium" for the citrus fruit type Oranges were transported along caravan routes from the orient to the Persian empire. The Moors brought them to Spain and used them medicinally and in religious services. The Spanish and Portuguese later introduced them into their new world territories. The treatment of survey done by orange juice, like lime juice, and the spread of oranges followed the paths of seafaring explorers who wanted to ensure having supplies of the fruit along their routes.

Nevertheless, up through the eighteenth and nineteenth centuries, sweet oranges were a delicacy reserved for the affluent. When it was discovered that oranges could be grown in temperate climates if protected from freezes, wealthy individuals began to grow them in glasshouses. The possession of such "orangeries" became a status symbol. Orange were first grown in Florida in 1565, but it was not until the United States took possession of the peninsula in 1820 that sweet oranges became in important U.S. commodity.

There are three main types of hybrid sweet oranges: bloods, normals, and navels. Blood possesses red bands in the pulp due to which it is unattractive to many Americans but they are popular in Europe. The most commonly grown normal type is the Valencia orange. Modern Valencia cultivars were introduced in California in 1876 and in Florida in 1877 by an Englishman who had become familiar with the variety in the Azores. Valencia oranges are now the most important variety grown in Florida. They produce a richly flavoured juice that sets the standard by which other orange juices are judged. Because we have come to expect orange juice to taste a certain way, commercial orange juice is often blended in order to ensure consistency in flavour and texture. In many cases, the juice is even filtered, freeze-dried, and reconstituted with controlled amounts of citrus peel and pulp. Large-scale navel orange production is a recent phenomenon, but a type of navel orange was known in Europe at least 300 years ago, and similar seedless oranges were either purposefully propagated or arose spontaneously in most orange-growing regions.

An abnormal small ovary is produced on the top of the regular ovary due to this phenomena navel oranges are formed. After maturity the second ovary becomes button or navel. Correlated with the aberrant production of the navel is sterility. The pollen produced by navel orange flowers is nonfunctional, and the ovules almost never produce viable seeds. New plants must be produced vegetatively, usually by grafting. Navel orange production in Florida began in 1835 and in California 3 years later. Navels are now the leading orange variety grown in California. Unlike the Valencias, navel oranges are primarily eating oranges because of their lack of seeds and he ease with which they can be peeled. The common English name "orange" is derived from the usual colour of the fruit, but fully ripened oranges are not necessarily orange.

In areas where temperatures never get cool (e.g., Thailand), oranges remain green to maturity. Cool temperatures thus contribute to the orange colour because they promote the release of carotene, an orange pigment, and the breakdown of chlorophyll. The fluctuation in temperature during ripening of orange changes the colour of fruits from orange to green and vice versa. Merchants, realizing the public desire for "orange" oranges, even though it makes no difference in their taste, will spray batches of green or blotchy oranges with ethylene to induce a colour change, or they may even resort to dyeing the fruits. Despite the modern preeminence of sweet oranges in the new world, they were not the first citrus grown in the Americas. Three others, the lemon, *Citrus limon*; the bitter orange, *C. aurantium*; and the citron, *C. medica* arrived almost 100 years earlier. The seeds of these three species are introduced by Columbus to the new world on his second voyage in 1943 and appears to have established them sucessfully on the island of Hispaniola. Bitter oranges are now used almost exclusively for the making of preserves, marmalade, and orange liqueurs.

The pulp of citrons is rarely eaten, but the peel is often candied and used in confections such as fruitcakes. Although native to southeastern Asia, citrons were known to the ancient Greeks, and Linnaeus used Pliny's name "citron" as the basis of the generic name for all of the "citrus" fruits. Citron seems to have given some of its genes in addition to its name to other citrus fruits. Taxonomic studies indicate that the citron figures in the ancestry of both the lemon and the lime. Like citrons and bitter oranges, lemons are practically never eaten alone, but their juice is used for flavouring everything from beverages to meat dishes.

Of all of the citrus fruits, lemons are most used for ancillary purposes. The oils present in lemon rinds are extracted and used for perfumes, clenaing products and deodorants. The inner portion of the rind is a valuable source of pectin. Morphologically the most distinctive member of the citrus group is *Citrus aurantiifolia*, the lime, originally cultivated in the Eat Indies. The Arabs were using limes by the tenth century A.D. and introduced them into Europe in the twelfth or thirteenth century. Women in the Renaissance French court carried limes with them for cosmetic purposes to whiten their skin and redden their lips. Lime cultivation in the New World began in the seventeenth century in the Spanish and Portuguese territories. After knowing the correlation between the consumption of fresh fruits juice and the absence of scurvy, British sailors were ordered to drink a daily ration of lime juice.

This juice can be kept well on long voyages, it is effective in small doses. It is a source of vitamin C which prevent scurvy.

We now know that it is the high vitamin C content of limes that prevent scurvy, but without knowing why, the British saved hundreds of lives with their kegs of lime juice. To this day, English sailors are known as "limeys." Grapefruits (*Citrus paradisi*) are thought to have originated as the result of a cross between a pummelo (*C. maxima*) and a sweet orange (*C. sinensis*) which spontaneously occurred in the West Indies, probably on Barbados. Thus, grapefruits can be considered a New World product and constitute a "species" that is only a few hundred years old. One of the ancestors, the pummelo, is little known in America today, but is relished in its native Thailand. Pummelos were apparently brought to the West Indies by the English or Dutch. In 1880 United States produced the grapefruits on commercial scale. Grapefruits with pink flesh arose as a bud mutation in a Florida plantation and were propagated by asexual grafting. The Texas "ruby reds" (with deep red-pink flesh) were developed in McAllen, Texas, in 1929.

The last important citrus fruit, *Citrus reticulata*, the tangerine, or Mandarin orange, was originally cultivated in China and is still an important fruit in the Far East. Recent numerical and chemical studies indicate that this species is a biological species and not a natural or artificial hybrid. Cultivation in the United States began in 1850, but the fruit has never enjoyed the popularity of oranges or grapefruits.

Squashes and their Relatives

In both the Old and the New World, members of the squash family (Cucurbitaceae) have figured in the early histories of man. Corn, beans, and squash, known as the "three sisters," were the mainstays of early agricultural Peoples of the southwestern United States, Mexico, and Peru. Squashes native of New World produced edible flesh and seeds. Seeds rich in sulphur containing amino acid. The blossoms are collected in some areas and used in soups. Fossilized edible squash remains from Peru have been dated to 5 to 6 thousand years before present. Undoubtedly, wild squashes were collected and eaten long before they were cultivated. In Africa, hunter-gather tribes such as the !Kung in the Kalahari Desert gather wild watermelons. Other melons figure in the earliest Egyptian writings. As already explained above, squashes are all

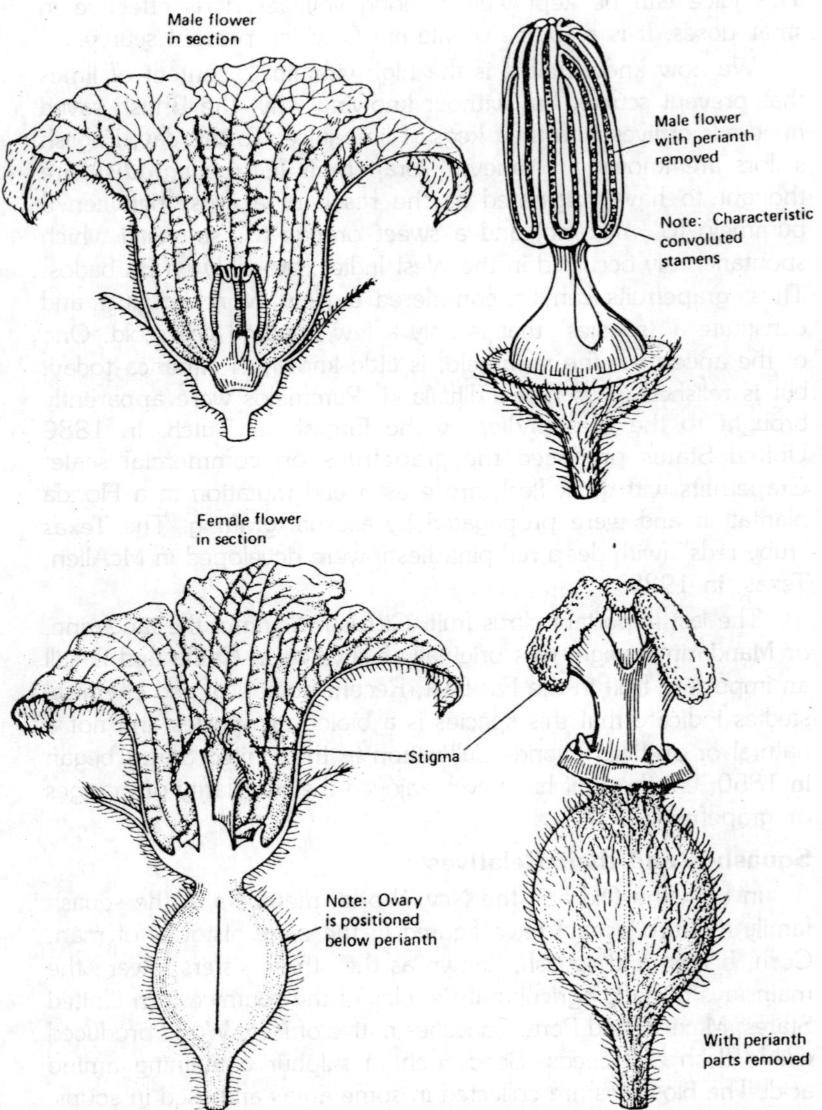

Fig. 5.2. Squash fruits are formed from inferior ovaries with the rind of the fruit a composite structure of the ovary wall and lower parts of the calyx and corolla.

native of New World and most of the melons are natives to Africa and Asia. Both melons and squashes are "squashes" in that belong to the squash family.

All cultivated members of the family are vines. Most are annuals, and the domesticated species discussed here are all monoecious. Since the plants are annuals, they are planted each year from seed. All members of this family have fruits called pepos produced from inferior ovaries. The skin, or rind, of a pepo is therefore composed of tissues of the ovary wall fused with the lower parts of the corolla. Because of their hard rids, many squashes and melons store well if not bruised or cut. The naming of true squashes members of the genus (*Cucurbita*) has been in a state of hopeless confusion for years, in part because common names of the edible squashes bear little relationship to the species from which they come.

There appear to be four widely cultivated species, three of which, *Cucurbita pepo*, *C. moschata*, and *C. maxima*, are commonly encountered in our supermarkets. All of these squashes are known only as domesticated species. Wild cucurbits usually have thin rinds and bitter flesh. *Cucurbita pepo*, initially domesticated in Mexico, is now the most widely cultivated and versatile species of squash in the United States. It provides us with Halloween pumpkins as well as yellow crookneck squashes, and, in immature form, zucchini and summer squash. Fossil remains believed to belong to this species have been uncovered in Mexico and are dated to be between 7,500 and 9,000 years old. Closely related to *Cucurbita pepo*, *C. moschata* has turned up in fossil deposits over 7,000 years old. Since the earliest of these fossils are from Mexico, this species also appears to have first been domesticated in Central America and subsequently taken to South America.

Many kinds of winter squashes including butternut and winter crookneck come from this species. Although *C. pepo* and *C. moschata* are naturally trailing vines, there has been selection for bush types in both species so that they take up less space in fields and are comparatively easy to tend and harvest. The last of the trio, *C. maxima*, has the least extensive fossil record of the commonly cultivated species. Peruvian material is the oldest evidence of its domestication. It is not found in ancient Central American deposits so the scientists believe its cultivation spread to North America from South America only after European. Nevertheless, *C. maxima* now provides us with hubbard, buttercup, and turban squashes. Several of its fruits forms are also called pumpkins.

In fact, since the flesh of winter squash is usually deeper orange in colour and less fibrous than that of *C. pepo*, it is often canned and sold as pumpkin pie filling. While the New World cucurbits provide important staple food crops, the Old World members provide luxuries. Except for the cucumber, all of the commercially exploited Old World members of the family tend to be bland or sweet and, in this country at least, most are eaten as dessert fruits. Watermelons (*Citrullus lanatus*), now grown extensively in California and Texas, are African natives that were appreciated by Europeans in ancient times. By the eleventh century A.D., the Chines were also growing them. Watermelons contains between 87 and 92 percent water. Sugar content in the watermelon is very high relative to that of other melons. The pulp of watermelon is acidic therefore it is not used as a flavouring for ice cream.

Other melons, varieties of *Cucumis melo*, seem also to be of African origin. References to melons do not appear in Egyptian or Greek writings but do occur in texts produced at the end of the Roman Empire. Selection eventually led to a wide array of different varieties including cantaloupe, and Persian, musk, Cranshaw, and honeydew melons. These sweet, refreshing melons may not seem very similar to cucumbers (*Cucumis sativus*), but they are actually closely related to them. Cucumbers are often eaten with sugar in their native region of southern Asia.

The long, narrow fruits might be considered the prototype of the modern thermos since they were often carried along on caravan journeys as a source of water. Egyptian slaves were reportedly provided with rations of leeks and garlic as protection against sunstroke, and cucumbers as a source of water. High water content and a bland flavour were prized qualities in the arid regions of Asia Minor and travelers introduced cucumbers throughout the Mediterranean region. Cucumbers are much more important item in the diet of peoples of western Asia than in the United States and are mentioned in the Bible. They are not widely use in United States. They are limited to use in salads and for pickle making. Several other members of the Cucurbitaceae are important in Latin America.

The white-flowered calabash, or bottle gourd, *Lagenaria siceraria*, has continued to baffle biologists interested in the origins of domesticated plants because it appears to have been cultivated in Ecuador and peru over 7,000 years ago and in Egypt over 3,000 years ago. How the species managed to be transported across either the Pacific or Atlantic Ocean before European contact has been the subject of many debates. Most workers now consider the disjunct distribution to be a natural one, attributable to the rafting of fruits across an ocean. Archaeological collections from Peru show a striking diversity of gourd shapes, several of which

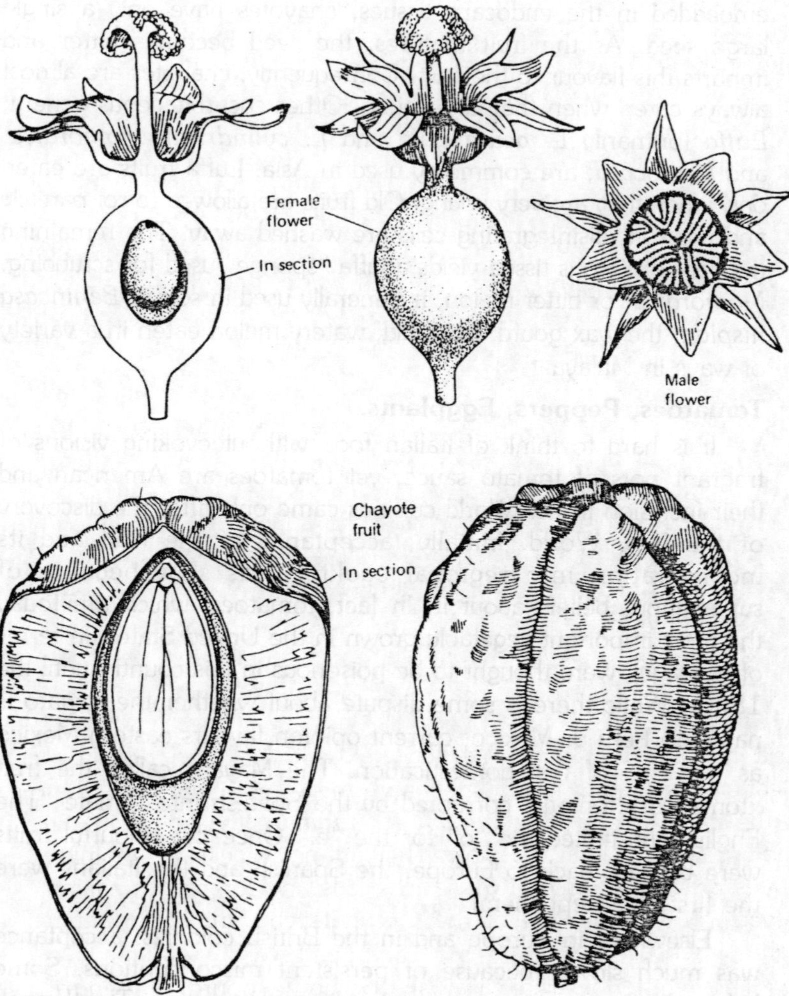

Fig. 5.3. Because each chayote contains a single large seed, entire fruits are planted to establish new vines.

had apparently been selected by Indians for specific functions. Gourds of a certain shape were used as *chicha* (corn beer) cups, while others served as water vessels and still others as bowls.

Various species of *Cucurbita* and the chayote all these were domesticated in Mexico in pre Columbian times. The fruit is green and pear-shaped and ocassionally seen in U.S. supermarkets. They can be boiled, fried, stuffed, or eaten in salads. Unlike most other members of the squash family which have numerous seeds embedded in the endocarp tissues, chayotes have only a single large seed. As the fruit matures, the seed becomes bitter and imparts this flavour to the fruit. Consequently, chayotes are almost always eaten when they are young. Other cucurbitaceous genera, *Luffa* (primarily *L. acutangula* and *L. cylindrica*), *Momordica*, and *Benincasa*, are commonly used in Asia. Luffa fruits are eaten only when they are very young. Old fruits are allowed to rot partially and the soft, disintegrating cells are washed away. The remaining network of fibrous tissue yields a luffa "sponge" used for scrubbing. *Momordica*, or bitter melon, is generally used in soups. *Benincasa hispida*, the wax gourd, is a mild, watery melon eaten in a variety of ways in Malaya.

Tomatoes, Peppers, Eggplants

It is hard to think of Italian food without evoking visions of fragrant pots of tomato sauce, yet tomatoes are American and their inclusion in Old World cuisines came only after the discovery of the New World. Initially, acceptance of the fruit and its incorporation into regional cooking was slow because of superstitious beliefs about it. In fact, tomatoes, which are today the most important vegetable grown in the United States (in terms of tonnage), were thought to be poisonous in this country until the 1800s. While there is some dispute about whether the tomato is native to Peru or Mexico, current opinion favours eastern Mexico as the area of first domestication. The Mayans called the fruit xtomatl or "tomatl," corrupted by the Spanish into tomate "The English substituted the "o" for the "e." Once the colourful fruits were brought back to Europe, the Spanish and the Italians were the first to accept them.

Elsewhere in Europe and in the British colonies, acceptance was much slower because of persistent misconceptions. Some believed that tomatoes had aphrodisiac properties. The French called tomatoes *pommes d' amour* (love apples), but this was apparently a misinterpretation of the Italian name *pomo d' oro* (meaning golden apples) or a variant of *pomi dei Moro*, a name that referred to the introduction of the fruit into Europe by the Moors. So many European members of tomatoes family have bitter fruits. The fruits containing toxic or hallucinogenic compounds. Therefore initially tomatoes were thought to be poisonous. The German common name "wolf-peach" reflected the belief that the fruits could be used in cabals to evoke werewolves. Linnaeus formalized this early appellation by giving the name *Lycopersicon esculentum*, Latin for "juicy wolf peach," to the species.

The British introduced tomatoes to temperate North America. Initially they grown these plants only as ornamental plants. Until

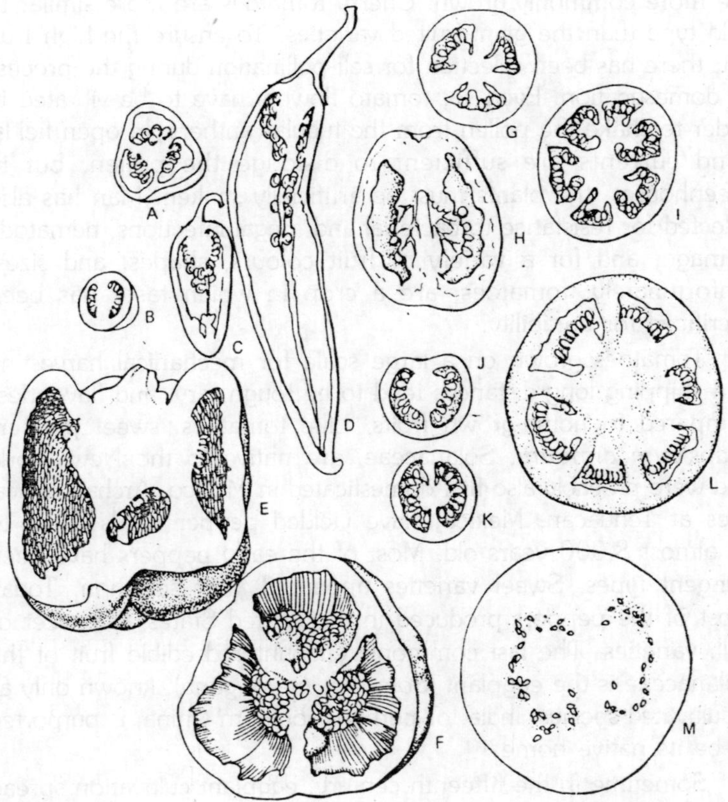

Fig. 5.4. Cross sections of fruits of the Solanaceae. A-D—hot peppers; E-F—sweet bell peppers; G-H—paste (plum) tomatoes; I-L—salad tomatoes; J-K—cherry tomatoes; M—Japanese eggplant.

1800, the fruits were relegated to a role as a pustule remover. According to an old farm journal, doubts about the fruit's edibility were finally dramatically removed in 1820 when Colonel Robert Gibbon Johnson announced that at noon on September 26, he would eat a bushel of the dreaded fruits. Two thousand people thought him mad and turned out to witness the event. To their astonishment, he survived. In the ensuing 2 years, tomatoes became a relatively popular fruit, but the spectacular surge in tomato cultivation came after 1920. Most of this increased productivity went into processed goods such as tomato juice, canned tomatoes, tomato paste, and catsup.

Wild tomatoes are outbreeding perennial herbs. Having small red berries which have two carpels. Today the giant-fruited varieties are more commonly grown. Cherry tomatoes are more similar to wild type than the giant-fruited varieties. To ensure the high fruit set, there has been selection for self-pollination during the process of domestication. Even so, tomato flowers have to be vibrated in order to shake the pollen from the tubular anthers. In open fields, wind currents are sufficient to dislodge the pollen, but in greenhouses, the plants must be artificially shaken. Man has also selected for resistance to bacterial and fungal infections, nematode damage, and for a variety of fruit colours, shapes, and sizes. Unfortunately, tomatoes, are a crop in which taste has been sacrificed for durability.

Tomatoes grown on a large scale for mechanical harvesting and shipping long distances tend to be tough, dry, and flavourless compared to home-grown fruits. Like tomatoes, sweet peppers (*Capsicum annuum*, Solanaceae, are native to the New World and were probably also first domesticated in Mexico. Archaeological sites at Tehuacan, Mexico, have yielded pepper seeds dated to be almost 8,000 years old. Most of the early peppers have been pungent types. Sweet varieties appeared after selection. Today most of the peppers produced in the United States are sweet or bell, varieties. The last commonly encountered edible fruit of the Solanaceae is the eggplant (*Solanum melongena*), known only as a cultivated species. India, or perhaps southern China, is purported to be its native home.

Sometime in the fifteenth century, eggplant cultivation spread to Europe and later to the New World. Eggplants have remained a very important dietary item in India but usually serve as an accessory food in other countries. In America there is a decrease in their popularity because of characteristic of eggplants are the browning of the flesh once the fruits are peeled or skinned and tendency towards bitterness. While the name eggplant may seem inappropriate for the large, ovoid, black-purple fruits that are now marketed, varieties common a few hundred years ago had small fruits that more closely resembled true eggs.

Coconuts

There is a South Seas proverb, "He who plants a coconut tree, plants food and drink, vessels and clothing, a habitation for himself, and a heritage for his children." Because of the versatility reflected in this adage, coconuts have earned the designation as the greatest provider in the tropics. Coconuts (*Cocosnucifera*, Arecaceae) can yield oil, fiber, food, and drink. Since we discuss oil production and fiber, we mention here only the history of the coconut and its use as a primary food. Discussing its origin is not, however, an easy task. Controversy has raged for years about the coconut's native home, because the fruits were present on the Pacific coasts of South America, southeast Asia, and Polynesia before Europeans reached the New World. Various geographers have suggested that ancient voyagers crossed the Pacific before 1492, but evidence currently supports the hypothesis that coconuts are native to the Indo-pacific region and that they dispersed across oceans passively by means of currents. Coconuts are now planted wherever the climate supports their growth.

Coconut plants are monocotyledons with a trunk composed of sheathing leaf bases. After germination and establishment the plants begin to flower and fruit, so it takes several years to flower and fruit. The species is monoecious. Each inflorescence bears numerous male and few female flowers. The fruit is formed from a flower with three carpels, only one of which develops. The mature fruit contains one seed, the largest known. The embryo itself is small and located near the stem end. Initially, the endosperm is a liquid containing free nuclei. This is the liquid drunk from green coconuts in many tropical countries. As the endosperm matures, cell walls form around the nuclei, and the endosperm solidifies into an oil-rich layer of coconut "meat" inside the seed coat. It is the solidified endosperm that we eat in pieces or grated in cookies, cakes, and pies.

Coconut "milk" is made by soaking grated coconut meat in water and squeezing out the liquid. If mature coconuts are not utilized or harmed, the embryo can germinate within the coconut since there is no dormancy period. The germinating seedling eventually extends the tip of the cotyledon through one of the eyes. The base of the embryo swells into an absorbing organ that eventually fills the entire cavity of the coconut as it digests the endosperm. The swollen organ, called a coconut apple, can also be eaten. In addition to providing coconut meat, or copra, for eating or oil production, the stalk can be fed to animals. The inflorescence stalk can also be tapped and he exuded sap used as a basis for fermentation. Occasionally, the growing tip of a coconut palm is removed and eaten as a vegetable, but since this results in the death of a valuable tree, it is rarely done. Hearts of palm are therefore usually obtained from other palm species.

Dates

The fruit of the date palm (*Phoenix dactylifera*, Arecaceae) were an important food in biblical times. Fruits of wild relatives of the modern commercial date were presumably gathered for thousands of years by tribes wandering across the arid regions surrounding the Mediterranean, but by about 4500 B.C., dates had bee domesticated. According to Muslims the date palm was made from the dust leaf after God made Adam. Dates would have seemed like a divine gift to desert travelers because they indicated the presence of water, which meant food and relief from the parching heat. They are highly nutritious: containing 75% carbohydrates and 2% protein. It is thus to surprising that the date palm was considered the "tree of life" to Bedouin people. Date palms are dioecious, but as early as 2300 B.C., agriculturalists had learned to hang a male inflorescence in a female tree to enhance pollination. Modern machines now blow pollen across the female flowers when the stigmas are receptive.

Pineapples

Pineapples (*Ananas comosus*, Bromeliaceae) are native to the New World. By the middle of fifteenth century they were widely cultivated by native people. During his second voyage Columbus described pineapples fruits and noted that they resembled pine cones, a similarity that led to the common English name. American Indians considered the pineapple a symbol of hospitality, and they used the sweet juice for making an alcoholic beverage and a poultice, and as a component of arrow poison concoctions. Pineapples are multiple fruits formed by the fused ovaries of 100

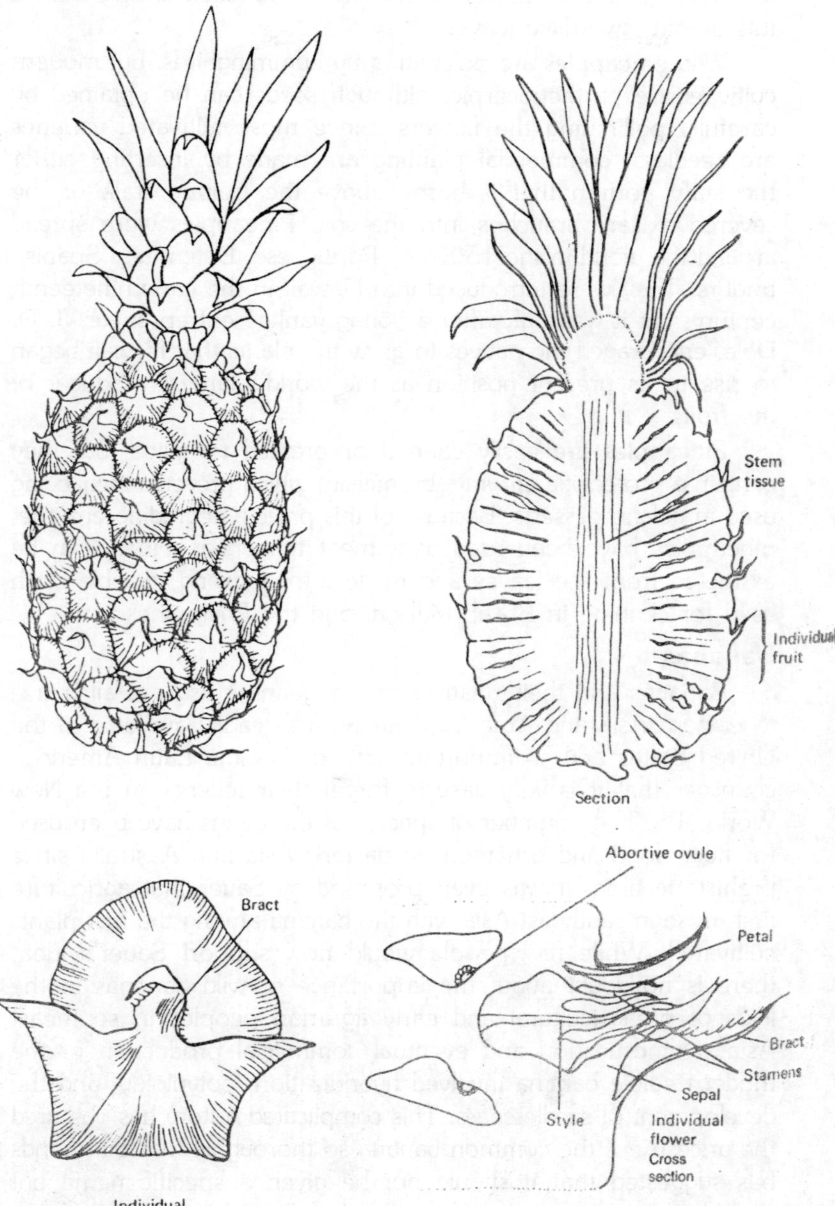

Fig. 5.5. A pineapple fruit is composed of 100 to 200 ovaries of separate flowers fused together and to the flowering stem.

to 200 independent flowers borne on a stout stalk arising from a tuft of stiff, swordlike leaves.

Wild pineapples are pollinating by hummingbirds, but modern cultivars are parthenocarpic, although seeds can be obtained by carefully pollinating the flowers. Since most cultivated varieties are seedless, commercial planting are made by inserting either the leafy portion that is borne above the fruiting stalk or the severed axillary branches into the soil. Pineapples were spread around the world in the 1500s by Portuguese, Dutch, and Spanish traders. They were introduced into Hawaii in the early nineteenth century, but it was only after a young yankee entrepreneur, J. D. Dole, encouraged the natives to grow the plants that Hawaii began to rise to its present position as the world's largest producer of the fruits.

Pineapples are easily canned or pressed for juice, but they contain a proteolytic enzyme, bromelain, which prevents their being used in gelatin desserts. Because of this protein-degrading enzyme, pineapples have been used as a meat tenderizer. Fibers can be extracted from the leaves and made into a strong, flexible cloth used for clothing in Brazil, Malaya, and the Philippines.

Bananas

Bananas and their relatives in the genus *Musa* are all native to eastern Asia and America. Banana are readily available in the United States and so important a food in some Latin American countries that it is very easy to forget their recency in the New World. The large number of species of the genus have been used for fiber, food and ornament in eastern Asia and Australia since prehistoric time. It was even proposed by Sauer that agriculture first arose in southeast Asia with the banana among the first plants cultivated. While few people would now support Sauer's idea, there is no doubt about the importance of wild bananas in the lives of preagricultural and early agrarian peoples in southeast Asia. Domestication and eventual commecial production of the modern edible banana involved hybridization, polyploidy, and the development of seedlessness. This complicated history has obscured the ancestry of the common banana so thoroughly that Simmonds has suggested that it should not be given a specific name but should merely be referred to as *Musa* followed by cultivar name. Other authors, however, still prefer to use Linnaeus's species name, *Musa paradisiaca* for sweet and starchy bananas. From their native southwestern Pacific home, bananas spread to India by 600 B.C. They were probably introduced independently into Africa and eastward across the Pacific.

In 1522, bananas were taken from the western coats of Africa to the Canary Islands as food for slaves being sent across the Atlantic. The first clones to occur in the New World were planted in the West Indies in the seventeenth century. Included in these early clones was at least one type of sweet banana, the "Silk Fig." Other early cultivars were starchy and were generally best when cooked before eating. While bananas are used for beer and are steamed, boiled, dried, or roasted in their native region of Asia, they are usually simply eaten fresh or fried in the New World.

The common eating bananas found throughout the world today belong primarily to the "Gros Michel" variety, a type derived within the last 200 years and introduced into Dominica in the early nineteenth century. The story of the banana's rise to a prominent agriculture position is intimately linked to the history of the United Fruit Company, which was formed in 1899 by the merger of the Boston Fruit Company with Keith Interests. Both of these earlier companies had been transporting bananas to the United States from Central America for 10 years before the merger. Yet, since bananas do not keep well, and neither company had a good distribution system, their operations remained small.

In 1990, the newly formed company developed a successful system of transport, and the United Fruit Company began to expand into Panama (1903), Guatemala (1907), E1 Salvador, and Costa Rica (in the 1920s). Nevertheless, it was 1929 before the first shipments of bananas were sent directly to the United States. In the 1930s and 1940s the company extended its operations into Colombia and Ecuador. Cargos of bananas were spoiled because carbon dioxide and ethylene produced by bruised and ripe banana induced ripening and caused rotting of entire shipments. Experiments eventually reflected that if bananas are kept at 100°C (56°F) and 90% humidity, ripening is delayed. Conversely, ethylene can be profitably used at a later time to induce ripening of green bananas.

Once the physiology of banana ripening was understood, shippers were able to prevent large-scale spoilage and even ship green bananas that they could ripen at will when they reached their destination. As a result, banana production boomed and the United Fruit Company continued to expand. For several decades, United Fruit essentially owned the land and controlled the workers on huge acreages of several Central and South American countries, causing them to be referred to as the "banana republics." By World War II, in the face of public pressure and political reorganization within these countries, the influence of the United Fruit Company began to wane.

Still, the importance of banana cultivation in many tropical American countries is reflected by the fact that bananas now constitute the major source of carbohydrate for millions of their people. Any vegetative parts of bananas gives rise to banana plant, corms, pieces of corms, or suckers. Adult plants are essentially giant herbs, although they are called trees and their cylindrical stalks commonly considered trunks. Lateral stem form at the base of the main stem producing suckers, most of which are removed in plantations.

Plants start to flower 2 to 6 months after initial shoot elongation. The flowers are borne on long, pendant, monoecious inflorescences with the female flowers at the base of the stalks and male flowers at the apex. Female flowers at the base of the stalks and made flowers at the apex. Female flowers are produced in clusters, each protected by a large bract. As a cluster of 12 to 20 female flowers becomes mature, the subtending bract peels back. Each cluster of female flowers produces a "bunch," or "hand," of bananas when it matures. In some regions, immature bananas are covered with plastic bags to prevent damage from soil, sun, dust, birds, or insects. Bananas are often cut by hand by two workers, one of whom cuts the inflorescence stalk, while the other stands ready to receive the heavy mass of fruit on his back. In this way, fruit damage is minimized. Care must be taken to avoid bruising or wounding the bananas, because such injuries cause the fruits to release ethylene. For shipment, only perfect bananas are used. They are picked green, washed, dipped in pesticide, and packed. The main banana stalk is usually cut at the base after fruiting. The bananas plant can be replanted each year or suckers left after the previous year's plant allowed to develop. If the initial planting was in rows, a sucker from the same side of each of the plants can be left to develop. The following year, the rows simply shift a few centimeters. This practice of allowing natural vegetative reproduction to produce subsequent plants can continue over a 5-to 25-year period. Bananas can yield impressive amounts of carbohydrate per unit area, with yields similar to those of potatoes (between 14 to 47 tons per acre).

Figs

Mulberries belongs to the family, Moraceae. Mulberries contains figs which an importants subtropical fruit. The fig genus *Ficus* contains over 800 species, many of which are cultivated, but only one of which, *Ficus carica*, produces a commercially important fruit. This fig, a native to the Mediterranean region, is mentioned in Egyptian documents (2700 B.C.), and is frequently referred to in the Bible. Adam and Eve were supposed to have sewed together fig leaves to cover themselves once they had learned to be ashamed of their naked bodies. To the Greeks, an edible fig was a "sykon," a designation that led to the modern term of syconium for the fruit, which is actually an inflorescence enclosed inside receptacular tissue. Wild figs are disseminated by birds or bats who eat the fruits and later defecate the tiny seeds.

People generally use cuttings for propagation of cultivated varieties. A few figs are either self-pollinating or parthenocarpic. The popular Kadota ad Mission figs, for example, can produce fruits without pollinators, but many wild figs, and most of the edible varieties, have a very elaborate pollination system involving tiny wasps that mate inside the fruits. Newly emerged and mated females carrying pollen from male flowers of their natal figs, fly to immature figs and enter. While laying eggs in modified female flowers of the immature figs, and females pollinate normal, fertile female flowers. Historically the most popular edible fig is Smyrna figs. They contain only female flowers in their syconia. In order to produce fruits, the flowers must be pollinated with pollen from male flowers in syconia on different trees. These pollen doner trees are known as "caprifigs". In the Mediterranean region, Smyrna and caprifigs naturally grow together. When Smyrna figs were brought to the New World, the pollination system was not fully appreciated and the inedible caprifigs were left behind. Naturally, the Smyrna trees failed to produce fruits. Eventually caprifig trees and the wasps were imported and successful American fig cultivation began.

Breadfruit and Jackfruit

Like bananas, breadfruits is an important source of carbohydrate for people in many tropical areas. A native of

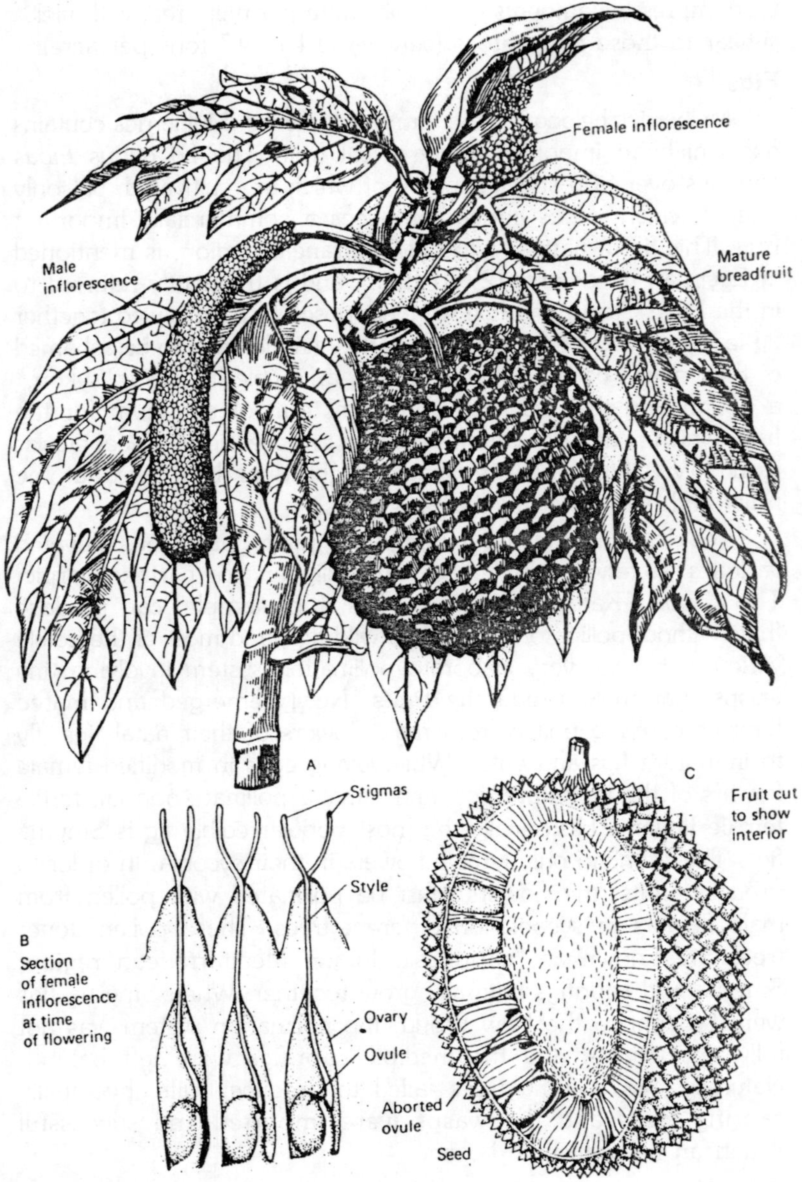

Fig. 5.6. The breadfruit fruit. A—Flower and flowering branch; B—Cross section of a portion of a female flower; C—Cross-section of part of a fruit.

Polynesia, breadfruit (*Artocarpus altilis*, Moraceae), has been cultivated since ancient times. Breadfruits are tall (to 20 m) trees that have compound, unisexual inflorescences. Female inflorescences develop into a round, multiple fruit about 10 to 30 cm in diameter. The moist, edible pulp is formed by abortive flowers. They few seeds that are produced are not normally eaten but can be used in times of food shortages. An average fruit usually contains abt 20 percent carbohydrates, and when cooked is similar in taste and consistency to a while potato. Early European explorers describing the plants as producing masses of bread. They were amazed when the first encountered breadfruits.

In an attempt to capitalize on this bounteous (and cheap) natural source of carbohydrates, Britain sent Captain Bligh and *H.M.S. Bounty* to Tahiti to collect breadfruits for introduction into the New World colonies. In 1773, Captain Bligh headed out from England with his ship equipped to be a floating greenhouse. It took Bligh and his men 6 months to collect seedling breadfruits and prepare them for the long journey to the West Indies. During their stay in Tahiti where they collected the plants, the crew began to savor the easy island life and compliant women. Soon after setting sail for the West Indies, they mutinied and set Bligh and 18 loyal seamen adrift in a longboat with a small cache of rations. Bligh survived the mutiny, crossing 3,618 nautical miles of the Pacific Ocean to Timor. Undaunted, he soon returned to Tahiti in another ship, *H.M.S. Providence*, and during his second voyage successfully established the trees in Jamaica. Today, they grow throughout the humid tropics. The morphology of Jackfruit (*Artocarpus heterophyllus*, Moraceae) is similar to the breadfruit, but its fruits contain a higher percentage (38%) of carbohydrates than its relative. It is commonly eaten as a desert fruit. Large-scale cultivation has tended to remain restricted to its native range of India and Sri Lanka.

Avocados

Avocados or alligator pears (*Persea americana*, Lauraceae, have a controversial history and seem to defy our often repeated theories about animal-dispersed fleshy fruits. Fleshy fruits consists primarily of water and sugars or in case of bananas and breadfruits, starch as well. Therefore fleshy fruits are low in calories. Avocados have a mesocarp that is extremely rich in oil.. Up to 30 percent of the pulp of cultivated varieties (on a dry weight basis) can be oil. As a result, avocados have the highest energy containing fruit pulp (from 2,000 to 2,800 calories per kilogram) known. Moreover, the seed is not protected in any way from the sharp teeth or digestive juices of an animal that might feed on the fruit.

It has been suggested that extinct large animals were the original dispersers of avocados, but the postulation of a large animal does not explain how a seed without a hard endocarp is protected in its journey from the mouth through the digestive system of such an animal. The natural dispersal of avocados thus remains a mystery. Avocados are known only as a cultivated species and yet they appear in some of the oldest archaeological deposits in southern Mexico (7000 B.C.).

It is possible that avocados were independently domesticated three times in the America, giving rise to the West Indian, Guatemalan and the Mexican (or Drymifolia) varieties. The Mexican variety has a small (about 250 gm, or 8 oz) fruit with a thin, smooth skin, a seed loose within the cavity, and a 30 percent oil content. Guatemalan fruits weight between 500 and 1000 g (1 to 2 lb) and have thick, warty skins They West Indian types are about the same size with thick but smooth skins and a mesocarp with only 8 to 10 percent oil. Avocados were first planted as a commercial venture in the United States in Florida in 1893, but acceptance was slow.

It is often said that the recent popularity of the avocado is a direct result of a calculated publicity campaign. Having failed to combat historical, adverse feelings about the fruits, growers (on the advice of an advertising agent) began loudly to deny that the fruits had any aphrodisiac properties. In the 1920s, few other claims would have been able to guarantee an immediate demand. The campaign was a success, and the popularity of avocados has been increasing ever since. The relationship between avocados and sex had, in fact, been claimed b native American Indians. The world avocado comes from the Aztec world "ahuacacuahatl" meaning testicle tree, which refers either to its stimulating properties or the appearance of the fruits that commonly hang in pairs. California and Florida now the big supplier to the country with an abundance of avocados during good fruiting years. Florida produces large, smooth skinned West Indian types while in California, most fruits are hybrids between the Mexican and Guatemalan.

Mangoes

Of all of the truly tropical fruits, mangoes (*Mangifera indica*, Anacardiaceae) are probably the most popular on a worldwide basis. They are particularly appreciated in their native region of southeast Asia where not only the pulp but also the seeds have been used for food. In times of emergency, the cotyledons of the seed a can be ground into a flourlike substitute and the leaves of the trees can be used for cattle food. Mangoes are naturally large trees and attains the height of 34 m. Where they are planted and irrigated in dry areas, the trees are much smaller which resemble orchards. Mangoes bears small flowers borne in long clusters. Usually only one flower per cluster develops a fruit. In full fruit, a mango tree looks as if someone has tied fruits on the branches because the fruits hang on long, bare stalks. The usually stringy pulp of a mango is the mesocarp that surrounds a stony endocarp enclosing the large seed.

Mangoes belong to the same family as poison ivy and, like many members of the family, produce a latex that is irritating to some individuals. Fortunately, latex production is restricted to the leathery skin, so peeling the fruit can generally eliminate the source of irritation. Portuguese brought mangoes to Brazil in the early 1700s. It was brought from South America to the West Indies in 1742. They constitute an important seasonal food source for the poor populaces of Jamaica and Haiti. Until recently, few were exported, but they are now commonly found in many U.S. supermarkets.

Okar

Although the native region of okra (*Abelmoschus esculentus*, Malvaceae) is still uncertain, it is thought to be either southwest or south-central Asia. The spread of the domesticated species must have been very early because okra was used in ancient Egypt and spread from there to the Far East and Europe. Okra first reached the New World in the 1600s, and in the 1700s it was adopted as a crop by French immigrants in Louisiana. They referred to the vegetable by its African name, "gumbo," and used it with such frequency that it became an integral part of Cajun cooking. Okra belongs mallow family with hibiscus-like flowers that produce capsules resembling elongated cotton bolls. Many people find the mucilaginous coating on okra seeds objectionable but the gum acts as a thickening agent in soups, stews, and gumbos. Okra has maintained a place in America agriculture because it can tolerate even the hottest southern months and still produce a substantial fruit crop at the end of the long growing season.

Pomegrantes

Pomegrante is native the Near East and Africa. They are subtropical fruits. In ancient world their importance was as much symbolic and artistic as nutritional. The Hebrews embellished their buildings with pomegranate motifs, and the globular fruits with their numerous juicy red seeds eventually became a symbol of fertility and abundance. This symbolism was carried to Asia where fruits were offered to wedding guests who were expected to throw them onto the floor of the honeymoon suite so forcefully that they shattered, scattering their seeds. This practice was believed to ensure fertility and a large number of offspring for the newlyweds.

The broadcasting of pomegranate seeds has another, more sobering, connotation. The French word for the fruit, "grenade," is applied to a hand-thrown weapon used extensively since the seventeenth century that scatters not seeds, but deadly metal fragments. Moors took pomegranates with them when they conquered Spain in 800 A.D. The seeds of pomegrantes are used for food. The tannin obtained from the fruits for leather. The fruit was introduced by Spanish to the New World many centuries later. Pomegrantes are now grown from the southern United States to northern Chile and Argentina. Having attractive flowers and fruits they are planted primarily as ornamentals. In addition to being eaten fresh, the seeds can be expressed to yield a juice. If sweetened with sugar, the juice becomes grenadine syrup.

Papayas

Papayas (*Carica papaya*, Caricaceae) are smooth-textured, musky-flavoured fruits native to Central America but little known in the United States until recently. In parts of Central and South America, they find their way into practically every meal during the fruiting season. The demand in the United States for papayas has not traditionally been for the fresh fruit but for papain, an enzyme extracted from the latex exuded when the skin of the fruit is scored.

Papain is very effective in breaking down proteins and thus forms the basis of commercial meat tenderizers. It is also used in chewing gum an cosmetics. Medicinally papain is effective for digestive ailments. Papaya plants are short lived, soft wooded trees. They have umbrella-shaped growth form. The flowers are borne along the stems in leaf axils. When the fruits are mature, the leaves have been shed, leaving clusters of the ovoid fruits emerging directly from the stem. Inside each fruit is a mass of slimy (each 0.5 cm in diameter) black seeds. For latex tapping, green fruits are used, beginning when they are 10 cm in diameter. The skin is cored diagonally in such a way that only the outer part of the exocarp is cut. The latex which oozes out is collected the next day. A single fruit can be tapped several times. The fruit can still be eaten after having been subjected to tapping.

A Sample of More Exotic Tropical Fruits

While many of the tropical fruits we mention here can be tasted fresh only where they are grown, they are of interest since some do occasionally appear in specialty food stores or even supermarkets. Within their native ranges, some are important foods for local peoples during the fruiting season. The sops, species of the genus *Annona* (Annonaceae), are American natives. Several, such as the sugar apple (*A. squamosa*), the custard apple or sweet sop (*A. reticulata*), and the cherimoya (*A. cherimola*), are eaten as fruits or used for juice throughout the Neotropics. The flowers of *Annona* species have numerous pistils. Consequently, annonas produce compound fruits that have a characteristic segmented appearance on the outside.

The custard apple is the most widely consumed of the annonas because selection has led to consistently high quality in the fruits and to seedlessness. The species also tolerates a wider range of climates than its relative. It has been introduced to Florida and is grown there on a very limited scale. To some, however, the cherimoya is the finest of all of the sops, but its growht is restricted to motane tropical regions between elevations of about 1,400 to 2,000 m.

Loquats (*Eriobotrya japonica*, Rosaceae) are native to eastern and southern Asia and are the few tropical rosaceous fruits. They are carried to the Mediterranean area and to the southern United States where they are planted mainly as ornamentals. The large shrubs are valued for their coarse-textured foliage, clusters of white flowers, and the orange, ovoid fruits that mature in the spring if winters are mild. The sweet flavour of loquats is best appreciated when the fruits carefully ripe, but slightly immature fruits are used in the preparation of jams, jellies, and liqueurs.

The cross section of carambola (*Averrhoa carambola*, Oxalidaceae) look like star-shaped. It is a tart tropical fruit. The

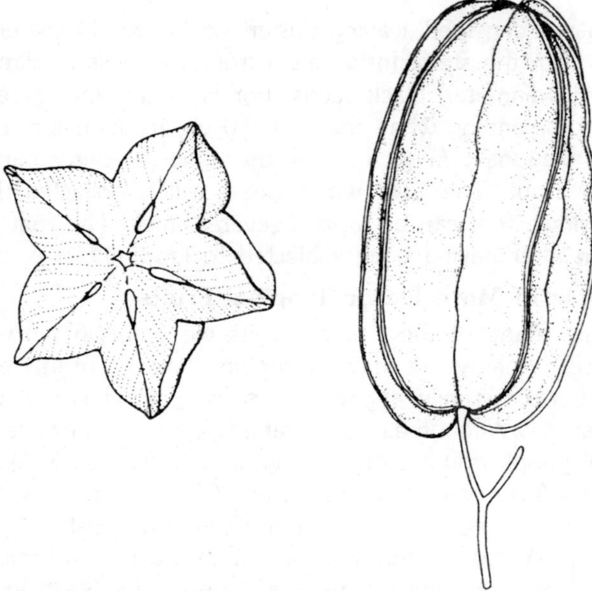

Fig. 5.7. The carambola develops from a five-carpellate ovary in such a way that the fruit is star shaped in cross-section.

flesh of carambola contains calcium oxalate crystals. These crystals dissolve in the saliva when the fruit is eaten and formed oxalic acid. The fruits are yellow and 3 to 6 cm long, and star shaped, reflecting the five carpels characteristic of the oxalis family. Native to Indonesia, star fruits are now planted throughout the tropics. Most Americans encounter litchis as canned fruits accompanying fortune cookies at the end of a Chinese meal. The so-called nuts, (*Litchi chinensis*, Sapindaceae), are natives of China. They are cultivated primarily in southern China and, to a limited extent, in subtropical areas of Florida. Litchis are produced on medium-sized trees (10 to 12 m tall). The edible portion of the warty, mauve-coloured, apricot-sized fruits is the aril that covers the seed. A recent introduction into U.S. supermarkets is the kiwi (*Actinidia chinensis*, Actinidiaceae).

Despite the fact that most Americans associate the fruit with New Zealand, it is a native of Asia. It has bee grown commercially in New Zealand and marketed as the Chinese gooseberry for 30 years. However, fruit sales remained poor until New Zealand held a national contest to rename the fruit. Under the name of kiwi, supplied by the contest winner, the fruit has steadily increased in popularity. Orchard of the dioecious vines now cover large acreages in New Zealand and a limited area of California. The pulp of the ovoid, fuzzy green fruits is delicate in flavour and blends well with other foods. In addition, slices of the fruit with their translucent, pale green flesh surrounding a narrow ring of tiny black seeds are striking and produce attractive desserts. Because of advertising for a canned fruit drink that includes their juice, passion fruits have become familiar to most Americans.

To many people, they seem to represent the essence of tropical fruits. It was the flowers, however, and not the fruits that originally attracted Europeans to passion flowers. The intricate and showy flower parts were considered by religious immigrants to represent the instruments of Christ's passion. Plants were consequently taken to Europe where they were grown as curiosities in green houses. Although the fruits had been eaten by native Americans long before European contact, systematic cultivation of passion flowers as a fruit crop began only a few hundred years ago.

There are 55 edible species. Out of these species, *Passiflora edulis* is now grown on the longest scale. This species is a native of Brazil and was introduced to Hawaii in 1810. These species grows best in tropical montane habitats at about 200 m elevation. Hawaii provides excellent growing conditions. In northwestern South American. *P. mollissima* is commonly grown. The parts of passion fruits that are eaten or pressed for juice are the arils surroundings the seeds. Each fruit has a tough pericarp surrounding numerous seeds, all of which are embedded in fleshy, aromatic, red-yellow arils. Since the plants are vine, plantations of passion fruits resemble vineyards. The perennial plants are evenly spaced and the branches trained along wires that are strung a meter or so above the ground.

Malay apples (*Syzygium malaccense*, Myrtaceae) and rose apples (*S. jambos*) are popular West Indian fruits. The species, introduced from southeastern Asia, have a tart, fresh, crisp flavour. The fruits are pear-shaped and are red or purple on the outside. Guava (*Psidium guajava*) is another myrtaceous fruit. It is a native of South America. From here spread to the Old World tropics. It is pear-shaped and yellow when ripe. It contain more vitamin C than most citrus fruits. Because guavas have a distinctly pungent taste when eaten raw, they are most commonly stewed or

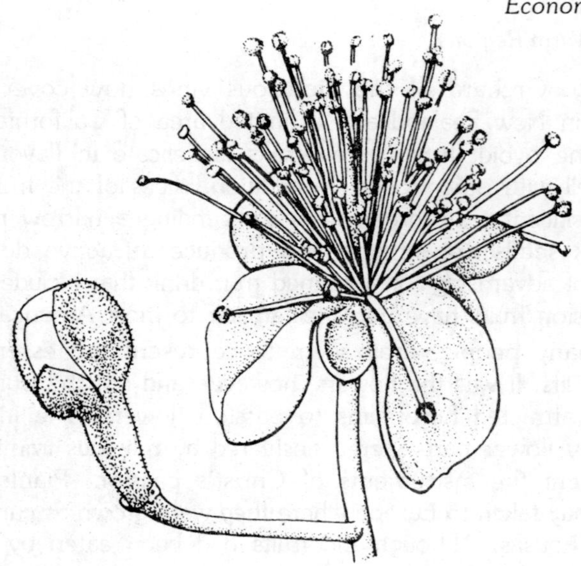

Fig. 5.8. Because they contain many small seeds and tend to be bitter, guavas are usually processed into jams and jellies.

processed into jams, jellies, or pastes. Sapote is a name give to an assemblage of species, the most common of which belong to the Sapotaceae. Two of these, the mammey sapote (*Pouteria sapota*) and the green sapote (*P. viride*) are both native American species that produce fruits about 5 to 20 cm (3 to 6 in) long with one to four seeds. The edible portion is the firm, reddish mesocarp surrounding the seed clusters. These two species are related to the more common sapodilla (*Manilkara zapota*), the fruit of the tree from which we obtain latex for chewing gum.

Tropical Nuts

Trees of temperate regions are main producer of our common nuts, several important nuts come from warm subtropical areas. But the most expensive, most delicious or tropical nut is macadamia. They are primarily produced in the Hawaii. In Hawaii they have been cultivated as ornamentals since the 1800s. Hawaii provides well-drained soils and abundant rainfall that are ideal for macadamia growth.

Commercial macadamia fruit production was initially hindered by the fact that macadamia trees to not produce a discrete, abundant fruit crop at one time of the year. In addition, the shells of the seeds are extremely difficult to crack. The taxonomy of macadamia was not understood for several years, and one of the bitter-seeded relatives of the sweet macadamia was occasionally grown by mistake or chance. Careful and deep study of the biology and morphology of the macadamia has now shown that the edible macadamias differ from the bitter types by many characters.. The problem of freeing macadamia nuts successfully from their shells was eased somewhat when it was discovered that drying the fruits before cracking facilitates the operation. Nevertheless, a hefty 300 lb/in^2 must still be exerted in order to crack the shells. While macadamia nut production is still limited by climatic conditions, asynchronous fruiting patterns, and tenacious shells, Hawaii now

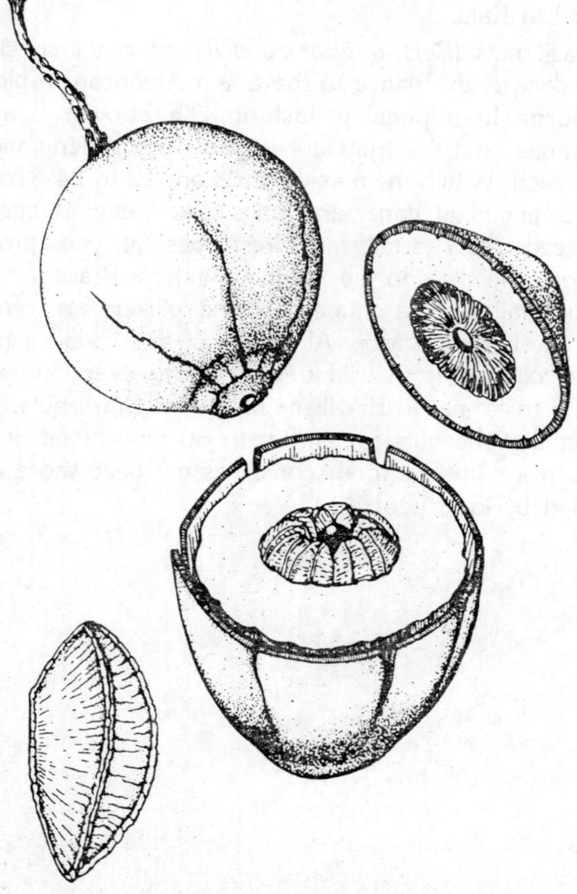

Fig. 5.9. Brazil nuts are the seeds of species of large forest trees.

enjoys a 6-months (July through January) season during which orchards yield an abundant harvest of the sweet, buttery nuts.

Cashews are native to forests of South America. They are popular nut in the United States. The cashews oils contains poisonous compounds in various parts of the plant. The latex in the seed coat is irritating but the seed itself is not. Cashews are produced on curious fruits. The "nut" is the embryo borne inside a hard seed coat. This structure sits atop a large, fleshy, inverted pear-shaped organ. Presumably, the fleshy portions attract animals that disperse the seeds. In Brazil, the fleshy portion of the fruit, called the apple, is occasionally crushed and fermented into a wine called Kaju.

Brazil nuts (*Bertholletia excelsa* and other species of the Lecythidaceae) are native to the South American tropics and are also borne in a peculiar fashion. The flower bear striking appearance, and the fruits are large woody spheres weight 1 to 2.5 kg each. Within the massive fruits are 12 to 24 wedge-shaped seeds arranged in concentric rings. Each seed is encased in a stony seedcoat. The tall, rain-forest trees can each produce over 300 pods. Needless to say, standing under a Brazil nut tree when the pods fall can be dangerous, and pickers are often injured during the harvest season. Almost all of the world's supply of the nuts is collected from wild trees. Despite their long association with the trees, native Brazilians rarely eat Brazil nuts. In view of the fact that the nuts contain up to 66 percent fat, it is strange that such a nutritious food source has not been more extensively exploited by local peoples.

6

FRUITS OF TEMPERATE REGIONS

Fruits, vegetables, grains and nuts are the categories given to edible plant parts in nutrition charts. Generally fruits can be defined as sweet plant products, generally eaten as dessert or as a first course and vetetables be those edible part eaten as entrees or side dishes, particularly seasoned with salt and pepper, according to a typical American shopper. But terminology in everyday use is quite different from botanical one. In botanical term, food commonly called as fruit and nuts, large proportion of starches and everyday vegetable are fruits such as green beans and cucumbers are fruits. Strawberries and blackberries are fruits not berries but tomatoes and eggplants are both fruits classified as berries according to a botanist.

In this chapter we are concerned with definition, classification of fruits. We will also discuss dessert fruits and nuts of temperate region and of warm regions in next chapter. Then, how can be explain a fruit, its formation and various kind?

A *fruit* is mature or ripened ovary with its parts and attached accessory structure, usually having mature ovule resulting from successful fertilization, known as seed and the polar nuclei.

Agamospermy

Seed production without fertilization. Mature seeds and fruits can be formed without fertilization. *Agamospermous* seeds contain only maternal genetic and cytoplasmic characters. However, such seeds' development can be stimulated by pollen without any contribution to them.

Parthenocarpy

Fruit maturation without seed formation is *parthenocarpy* and such seedless fruits are called *parthenocarpic*. Many parthenocarpic mutants for seedless fruit that mature without fertilization spontaneously.

A fruit is classified by the number of ovaries which produced it (one or several), by the position of the ovary (superior or inferior), by whether it is fleshy or dry at maturity, and by the way it releases its seeds (dehiscence). Anatomically, an ovary can be simple or compound. In the first case, it has one carpel, in the latter, two or more carpels. The notion of simple or compound and the concept of the carpel comes from an hypothesis called the *foliar theory* of the carpel.

According to this theory, the ovary was derived from a leaflike structure on which ovules were borne. During the course of evolution, this leafy structure was supposed to have folded, producing the closed ovary characteristic of the angiosperms. Each simple, folded, leaflike unit is considered to be a single carpel. A simple ovary is derived from one such structure and therefore has on carpel. A compound ovary is one that appears (usually fro anatomical study) to have been produced by the fusion of two or more of these leafy structure. The number of carpels in a compound ovary is therefore equal to the number of ancestral

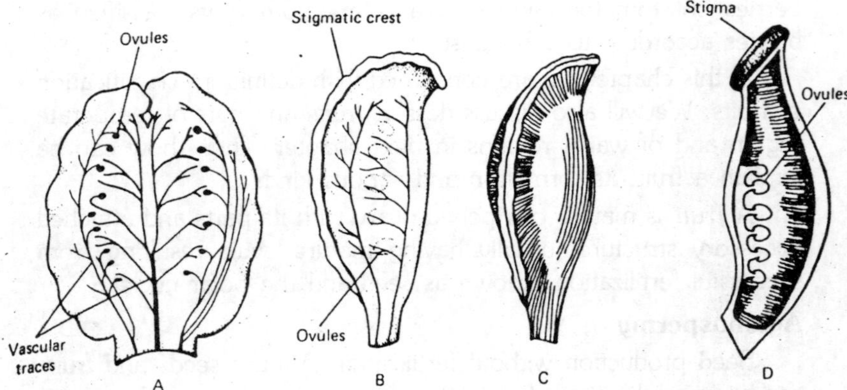

Fig. 6.1. It is postulated that the carpel was derived from the folding of a leaflike structure on which the ovules were brone (A). The modern carpel of Drimys winteri is believed to have resembled this early folded leaflike structure. (B) The vascular traces inside the Drimys carpel, (C) an exterior view, (D) cut away view.

leaflike structures fused into the pistil. While this explanation of the carpels' origin is not universally accepted virtually all plant descriptions use the number of carpels of an ovary as one of the diagnostic characters of a species.

A *simple fruit* derived from a single simple or compound ovary having one or more seeds surrounded by a pericarp, which is made up of three layers the endocarp, mesocarp, and exocarp. The last three terms refer to the inner (endo-), middle (meso-, and outer (exo-) layers of the ovary wall. The pericarp, with surrounding tissues (if present), can develop in various ways. Assortment of fruit is types characteristic of different species of angiosperms. For example, the pericarps can remain thin and become dry during fruit maturation. Thus formed dry fruits can be shed as a whole units, or can release their seeds. In the first case, the fruits are said to be dry and *indehiscent* (nonsplitting), e.g. ash and elm fruits and in the second case, dry and *dehiscent* (splitting) e.g. granium. Dry fruits, or their seeds, are generally dispersed by wind, water, or gravity.

Fig. 6.2. The fusion of several carpels is presumed to have led to a compound ovary.

Grains, or caryopses are important kind of indehiscent fruit with ovary wall having fused seed coat along. Legumes grouped under dry, dehiscent fruits that split along the two opposite margins of the pod. Fleshy nature of pericarp increases as the fruits matures in many cases. An enlargement of any one (or all) of the pericarp layers or a swelling

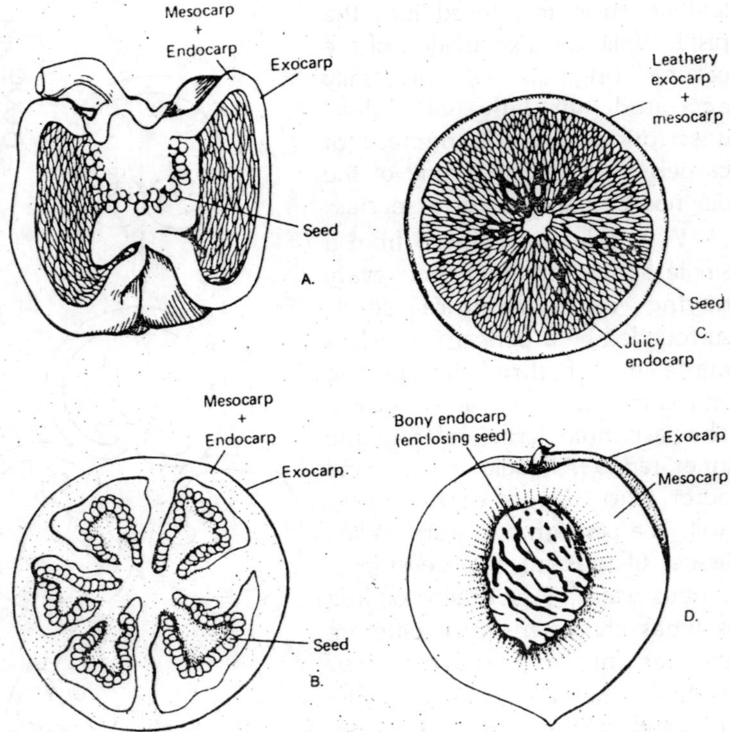

Fig. 6.3. The exocarp, mesocarp, and endocarp are represented in different ways. A—Slightly expanded mesocarp gives rise to the fleshy rind of the pepper. B—A tomato also has a relatively fleshy mesocarp and an elaboration of the tissues on which the seeds are borne. C—The fleshy part of an orange is made up of swollen hairs on the inside of the endocarp. D—A greatly expanded mesocarp produces the sweet flesh of a peach.

of accessory tissues can form a fleshy portion. If pericarp form fleshy portion a *berry* is formed and fruits is produced by a single superior ovary. A superior ovary of the floral parts. Berries can have one or several seeds, but *drupes* are the seeds without hard endocarp even *Citrus fruits* are *hesperidium*. Kind of berries having leathery rinds dotted with packets of oil and inner wall of endocarp having particular fleshy projecting hairs.

An ovary in inferior if present below the attachment of the floral parts (sepals, petals etc.), it is called *inferior*. Bases of the floral parts surrounding them is fused with wall of such ovary. The fruit wall therefore consists not produced by inferior ovaries are accessory fruits, but they are usually called by the names of the simple fruits they resembles. Coffee berries blueberries, and bananas are accessory fruits, but they are generally simple referred to as berries. *Pepos* is the special name given to the squash family, type of accessory fruits as they are economic important and have distinctive hard rind. In some cases, the perianth parts are fused at the base forming floral cup around the ovary. This cup can swell as the fruit matures and provide the majority of the fruit pulp. *Pomes* are the rose family formed in this way.

Aggregate fruits are formed when numerous, simple, superior ovaries borne within single flower, each matures to form, its part. Strawberries, raspberries, and blackberries are therefore aggregate fruits and not true berreis.

Multiple fruits are mostly tropical and are formed by the tight packing of fleshy fruits from many separate flowers. Fruits such as pineapples and figs are produced in this way. In other words multiple fruits are false fruits since each is actually composed of many fruits. Each segment of a pineapple, or each tiny achene inside a fig is a complete fruit.

Attraction of animals who will eat them but spit out is the main aim of fleshy fruits or it can be a nutrition a plant has as few resources in fruit pulp for the embryo of later development of seedling. Being the immediate products of photosynthesis sugars are less "expansive" for a plant than more complex compounds that require further synthesis (e.g. oils) or that contain limiting nutrients such as nitrogen (e.g. proteins or alkaloids). Therefore fleshy fruits parts tend to consist primarily of flavoured sugar solutions held in a cellulose matrix. Fruits such as the avocado and olive are exceptions in which fleshy fruits are watery, low in calories, and good sources of water-soluble vitamins.

In contrast to the pulp of fleshy fruits, seeds contain nutrients that are used by the embryo for emergence and establishment. Seeds therefore contains sufficient quantities of high-energy material such as fats, oils, and starches. Consequently, many plants have seeds that are protected by hard shells, seed coats, or toxic compounds from predators that destructively consumes seeds reducing its reproductive potential. Humans have certain advantage as they have learned to overcomes such derives for seed protection.

Apples and Their Relatives

Apples, the most important temperate fruit tree crop in the world belong to family Rosaceae and Rose family. Apples with

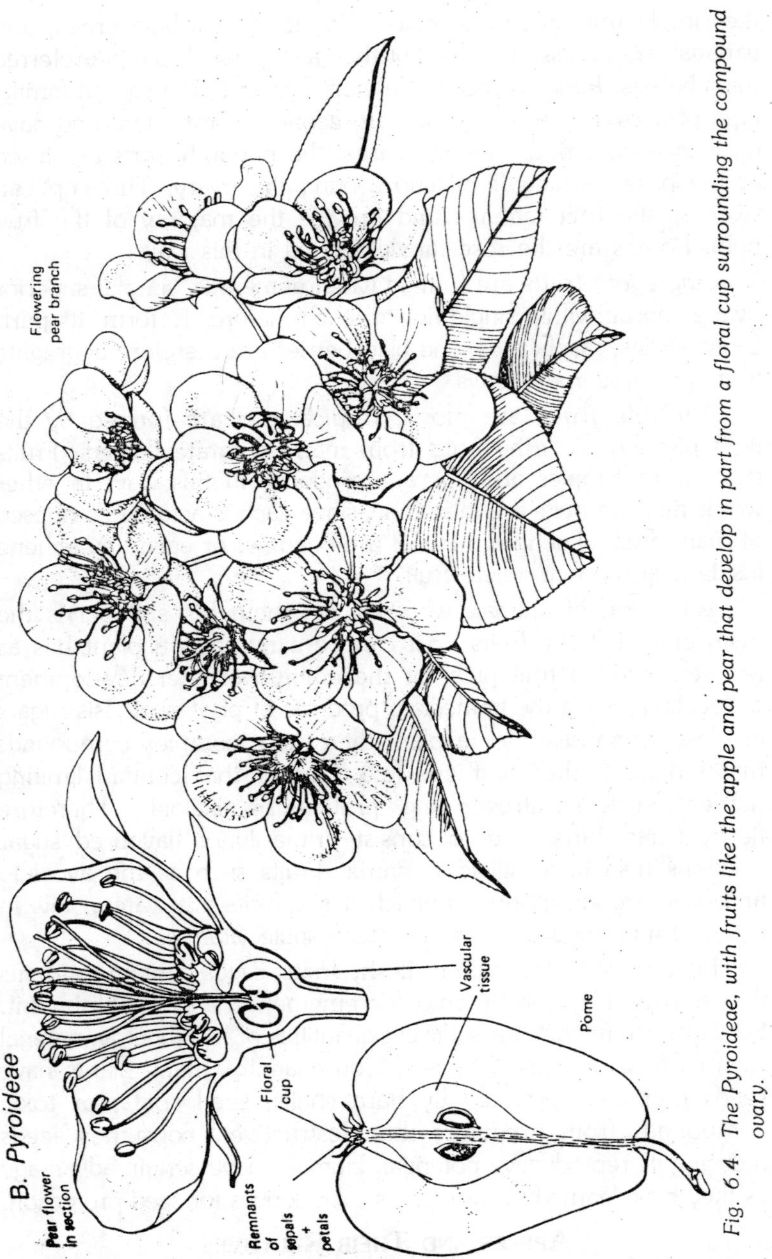

Fig. 6.4. The Pyroideae, with fruits like the apple and pear that develop in part from a floral cup surrounding the compound ovary.

their relatives, pears, peaches, plums, cherries, strawberries, and raspberries, all are grouped the same family, the Rosaceae, or rose family. However, because of the differences in between, they placed under separate subfamily from plums and peaches. Strawberries and raspberries are put in a third group. Most botanist divided the Rosaceae into four subfamilies, the Spiraeoideae, Pyroideae, Prunoideae and Rosoideae. The last three being important sources of edible fruits. The Pyroideae (also called the Pomoideae) produces pome fruits. Species in Prunoides produce drupes as fruits, and plants in the Rosoideae produce aggregate fruits.

Pyroideae

The Pyroideae contains two important genera of fruit plants, *Malus*, apple, and *Pyrus*, pear, as well as the less well known quince, *Cydonia*. Apples account for over 50 percent of the world's deciduous fruit tree production. *Malus pumila* is the species under which cultivated apples are usually placed. It is believed that the genus originated in western Asia but spread to North America long before any human arrived on the continent. Hence our cultivated, edible apples are natives of the Old World, despite the expression despite the expression "as American as apple pie". Apples have been eaten in Eurasia since prehistoric times, and selection must have been practiced very early since the Romans are known to have had at least 22 apple varieties. It shows that enough varieties had been selected by 1670, the Grand Duke of Tuscany gave a banquet in which he served 56 varieties of apples.

Today, more than 6,500 horticultural varieties have been registered. Unfortunately, in modern markets we seem to see fewer varieties than our grandfathers. They can probably list many kinds of apples—Cortlands, MacIntosh, Granny Smiths, Jonathans, Winesaps, Gravensteins, and Pippins (among others)—that they used to see regularly. Now, a few varieties, especially the Red and Yellow Delicious, developed in the United States, dominate in world trade, although two or three others can commonly be found. Apples normally have an ovary with five carpels, but one or more often contain no seeds. The edible fleshy part of the apple is the floral cup which expands as the fruit matures. The wall of the ovary is the thin, papery layer surrounding the core. Most of the apples grown today are triploids. Because the initial production of these triploid hybrids is time-consuming and costly, superior varieties, that have been developed are perpetuated by grafting.

Grafting also permits the use of a variety of rootstocks which have different adaptations for hardiness or pest resistance and which are therefore better suited to a variety of localities. The apple has been a part of folklore since ancient times. In Greek mythology, one of Hercules's tasks was to obtain the golden apples of Hesperides which were well guarded because they were supposed to provide immortality. In the book of Genesis, Eve was tempted by a serpent and took a bite of a forbidden fruit which most people have assumed was an apple. The generic name of the apple, *Malus* is after the Latin "malus" meaning bad, as it was presumed to be involved in the downfall of man. There is some dispute, however, about the identity of the golden apples of mythology or the fruit in the Garden of Eden because some authorities believes that apples did not grow in the Near-East or around the Mediterranean Sea when these episodes were supposed to have occurred.

According to this view, a late climatic change allowed the movement of wild apples into this region. To these workers, apricots are more likely the fruit in the stories rather than apples. Apricots are called golden apples today in the Near East and were the most abundant fruit in Palestine (next to figs) in early historical times. Still, most authorities believe apples to be native to the Middle East and do not find any reason to doubt that they have occured in early legends. There are also more recent tales involving apples, William Tell, a Swiss hero, is said to have been forced to shoot an apple from his son's head because he refused to pay homage to Austrian conquerors. It is believed that Newton discovered gravity when an apple fell on his head. In America, the legend of Johnny Appleseed, based on the life of John Chapman (1774–1845), recounts the migration of the professional arborist from Massachusetts as he trekked westward planting his beloved fruit trees. Pears are the second largest deciduous tree fruit crop on a worldwide basis, but less popular in United States.

Many species of pears appear to hybridize freely and to be involved in the production of the most widely cultivated species, *Pyrus communis*. Different opinions are there whether pear was originated in eastern Europe or primarily originated in western China like the apples. Both areas are regions of high diversity of *Pyrus* species. Unlike apples, there are no pears native to the New World. Either the English or the Spanish have introduced

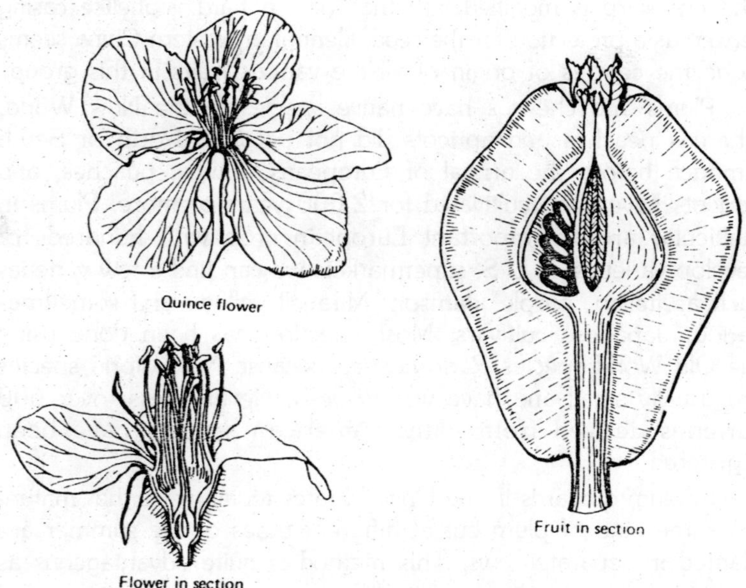

Fig. 6.5. Quince trees can reach 15 to 20 feet in height and are among the first plants to bloom in the spring.

the fruit to America. In Eurasia, pears have been cultivated since at least classical Grecian times Selection in pears have been primarily for different fruit appearances as main basis for selection in peas. Propagation is generally by means of grafting. The slightly gritty texture of pears is caused by the presence of stone cells, short, squarish cells with thick cell walls. These hard cells are present throughout the flesh of the fruit. The third economically important fruit of the Pyroideae is the quince (*Cydonia oblonga*). Quinces are sometimes used as rootstocks for pears, and like pears are more popular in Europe than in the United States. Quince fruits tend to have a hard flesh with noticeably more stone than pears. Consequently most quinces are made into preserves and conserves.

Prunoideae

Prunoideae, contain are several important fruit species all belonging to the same genus, *Prunus*. These are plums (primarily *P. domestica*), cherries (*P. avium* and *P. cerasus*), peaches (*P. persica*), and apricots (*P. armeniaca*). Fleshes mesocarp is the edible portion in all these fruits and the exocarp being the "skin".

The endocarp is modified into the "pit," a hard, stonelike casing serving as a protection for the seed. Central or western China seems to be the centres of origin of all the various fruits in this group.

Plums and cherries have native species in the New World, whereas peaches and apricots did not occur in North or South America before the arrival of Europeans. Plums, peaches, and apricots have been cultivated for 2,000 years or more. Plums in particular are an important European crop with hundreds of develop varieties. In U.S. supermarkets we can find a few varieties such as Italian, purple, damson, Mirabell yellow, and sometimes various Japanese cultivars. Most selection has been done using the Old World species *P. domestica* because New World species are usually small and have watery flesh. Nevertheless, over 600 varieties derived from native American species have been registered.

In plum orchards in the United States today, trees that mature the same kind of plum but at different times of the summer are planted in separate rows. This method is quite advantageous as there is no need to alter any procedure or machinery for prolonged harvest. Plums are cultivated in the United States in every state except Alaska, but the primary area of production is along the west coast. Prunes are plums which been dried without allowing them to ferment. Despite the common modern association of prunes and a laxative effect, their production dates back to times before modern methods of preservation because once dried, they can be kept for long periods without spoiling. Like plums, there are New World cherries, but cultivated cherries are essentially all derived from Old World species. Their cultivation in the United Stage began with the early settlers, 1847 is the most important date for the U.S. cherry industry, when one of the two brothers named Lewelling, settled in Oregon after a 300-mile oxcart trip from Iowa with several hundred cherry seedlings, eventually selected both the bing cherry and Lamberg.

Cherry is being the most important cultivar in the United States today, and the Lambert is the secondary most important U.S. variety. Cherries have a special meaning for Americans because of their association with young George Washington. A famous story tells that once Washington chopped down the cherry tree and confessed the deed to his father. But the toll is, unfortunately, fictitious. Parson Weems created the story for the fifth edition

(1806) of his book *The Life and Memorable Actions of George Washington*. Weems seems to have been more concerned with providing a good example of honesty than an accurate account. Today, cherry trees which are cultivated for their flowers rather than their fruits adorn the Tidal Basin of Washington, D.C. Drawing the visitors from all over the country, with a display that signal the arrival of spring, these trees are 'gift of Japanese government'.

Cherries are grouped according to taste into sweet and sour types. Sweet cherries are usually considered to belong to *Prunus avium* and sour types to *P. cerasus*. Within each of these groups, cherries are classified on the basis of the colour and firmness of their flesh. Maraschino cherries are made from sweet cherries, usually the Royal Anne variety, which have been bleached, deseeded, and soaked in a sugar solution to which red food colouring and flavouring have been added.

The production of maraschino cherries dates several hundred years to Italy where a cordial maraschino (named after the marasca cherry most commonly used in its preparation), was produced. The French later developed the sweet type of maraschino that has now replaced the original bitter Italian liqueur. Peaches are the third most important fruit crop in the United States, lagging behind oranges and apples. They are native to China where they have been cultivated for thousands of years and where they symbolize long life. In 1500s they were introduced to Mexico by Spaniards. Since that time, their cultivation in the New World has spread, and many selections have been made that allow their growth in a variety of climates. The United States is a leading peach produce.

Peaches are an important crop in California, Arkansas, Texas, Georgia, and New Jersey. Peaches fall into two major classes, freestones and clings. Freestones are peaches that have pits which separate easily from the flesh. The mesocarp of cling peaches adheres firmly to the pit. Like apples and plums, peaches are generally propagated by grafting, once a superior cultivar has been selected or discovered. Becoming increasingly popular, they are without their furry coats. Nectarines are descended from a mutant that was noted by a nurseryman and subsequently propagated by grafting. Apricots were first brought to Greece from Persia by Alexander the Great and to the New World by Spaniards who established orchards in California in 1770.

In ancient times, they figured prominently in folklore. In Persia, apricots were called "seeds of the sun." Chinese legends ascribed prophetic powers to them. Confucius is said to have worked out his philosophy under an apricot tree, and the indiscretions of the Duchess of Malfi in John Webster's English blood tragedy were uncovered when she ate apricots and claimed that they were bitter, a sign that she was pregnant. Apricots have also been associated with healthiness. In Hunza, a small kingdom in the Himalayas, apricots are an important part of the diet. The surprising long life and the robust health of the people in this tiny country is said to be due to apricots.

Rosoideae

The last important subfamily of the Rosaceae, the Rosoideae, contains the false berries of the fruit trade, strawberries, raspberries, and blackberries. Although all of these fruits contain many separate mature ovaries from a single flower their fleshy portions have different origins . Most of a strawberry is the swollen, convex receptacle on which the ovaries were borne. Strawberry is detached under the receptacle when picked. The fruits are tiny achenes that dot the surface of the swollen portion.

Blackberries and raspberries contain a mass of swollen ovaries. The receptacles of these fruits are not swollen and remain on the plant when they are picked. Thus, these berries are hollow. Each bump on a raspberry or blackberry is an individual, fleshy, mature ovary with a seed inside. Strawberries are one of the most beautiful fruits in the Rosaceae and the most important commercial fruit crop in he Rosoideae. Strawberries are increasingly popular fruits in American markets due to their delicate, yet distinctive flavour, their beautiful colour, shape, and odor, and their ability to combine with a variety of foods. The generic name *Fragaria* is derived from the Latin "fragrans" because of the fruit's delightful fragrance. The Anglo-Saxon name strawberry probably refers to the plant's habit of sending out runners, or "strewing" itself across the ground.

Strawberries occur wild in Eurasia and the Americas. The fruits of all species are edible and were eaten by native peoples of both the Old and New Worlds. Romans relished strawberries, and, for the farmers during the Middle Ages, they were one of the few delicacies available. European strawberries are small, however, and few berries are produced per plant. As North and South America were explored, new species of strawberries were found and sent to Europe. One such species was the Virginia strawberry, *F. virginiana*, found on the North American east coast. Other strawberries were found in Canada, and, in 1646, an exceptional species, *F. chiloensis*, was located in Chile. F. ananassa, the most common cultivated species is produced about 1750 as result of a spontaneous cross between the Virginia and the Chilean species in a European garden.

Today, strawberry cultivars have been developed that yield more fruit per unit per unit area per time than any other temperate dessert fruit. Many people simply lump raspberries and blackberries together as "brambles," but both actually consist of a number of species of the genus *Rubus* that occur natively in North America and Europe. The principal raspberries grown commercially are *R. idaeus* (red), a European species, and *R. occidentalis* (black), a North American taxon.

Black raspberries are grown in the Pacific Northwest as a source of purple dye and rarely consumed as fresh fruits. The dye, used to label meats with the familiar USDA ratings of prime, choice, etc., is completely edible since it is nothing more than blackberry juice. A cross between an octoploid California blackberry and a European raspberry first found in the California garden of Joshua Logan gave rise to the loganberry, a popular fruit in the western United States used for pies and jams.

CURRANTS AND GOOSEBERRIES

Both currants and gooseberries belong to he same genus, *Ribes* (Grossulariaceae). Their generic name is derived from the Arabic word *Rheum* for rhubarb, referring to a plant that produces an acid juice. There is a pronounced tartness, therefore all are better when cooked and used in jams, jellies, or pies. Northern Eurasia and North America have a numerous species of currants and gosseberries occurring natively. While the New World members have been eaten locally and used for breeding purposes, the most commonly grown species of currants and gooseberries are derived from European taxa.

Currants and gooseberries are accessory berries borne on long-lived shrubs. Mechanical harvesting is difficult for currant because of their sprawling habit. Gooseberry bushes are erect, but often heavily armed with spines. Currants are small (3 to 6 mm), red or black berries with smooth skins. The most commonly cultivated black currant is the European black currant (*R. nigrum*), having

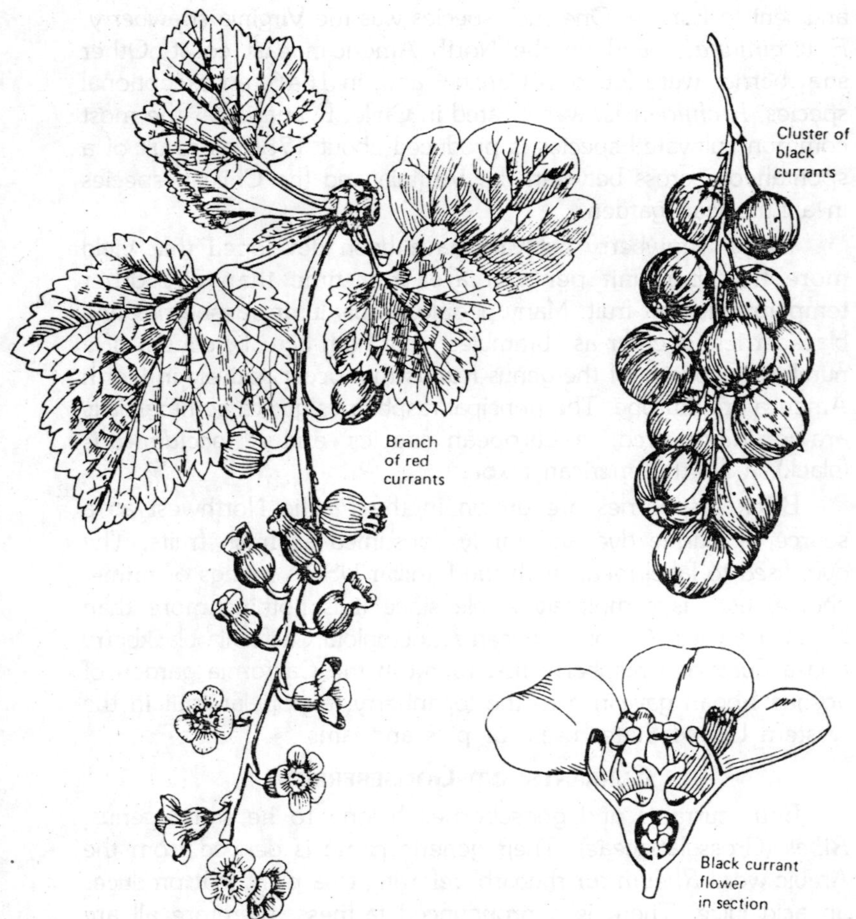

Fig. 6.6. Currant flowers are produced in pendulous racemes.

an especially strong, smoky flavour which gets better when combined with sugar in jams or liqueurs such as creme de cassis.

The red "cherry" currants (*R. sativum*) are generally used to make jam or are dried. In dried form, currants resemble small raisins and are often substituted for them in cookies, fruitcakes, and puddings. Because currants serve as a host for the pine blister rust (*Cronartium rubicola*), their cultivation is banned in the United States where pine trees are grown commercially. Currants have increased to the fungus due to breeding, but currant cultivation in the United States is still much lower than in Europe. The berries are green, translucent, and marked with stripes. Because they are exceedingly tart and rather bland, they are used exclusively in pies and jams. Gooseberries (primarily *Ribes uva-crispa*) have apparently been cultivated commercially for only about 100 years, although they were grown on small farms for at least a century or more before. Like currants, most of the breeding and cultivation of gooseberries have been in Europe. In the United States they are grown commercially only in Oregon.

Blueberries And Cranberries

Although blueberries and cranberries are not important worldwide, both are popular in the United States. Blueberry pie is almost as American as apple pie, and cranberries are an essential part of a Thanksgiving meal. Both of these berries belong to the Ericaceae, or health family. The family is probably best known for its ornamental shrubs: azaleas, rhododendrons, and mountain laurels. Like these ornamentals, cranberries and blueberries thrive best in acid soil. This is why blueberry cultivation is limited. But it not the only reason, also because the berries are difficult to pick.

Two types, high-bush blueberries, *Vaccinium corymbosum*, and low-bush blueberries, *V. angustifolium*, yield commercial fruit crops. Both species are native to North America and have been introduced into Eurasia. Cranberries are usually placed in the same genus as blue berries, a separate genus *Oxycoccus* is given to them by some botanists. Two species—*Vaccinium macrocarpon*, of North America, and *V. oxyoccos*, a widespread north temperate species—are cultivated. Cranberry plants are creeping shrubs that are grown in boggy areas which can be periodically flooded. Preventing of winter desiccation, protection of plants from freezing, and regulation of fruiting times and harvesting are the main purpose of flooding. The ripe berries are shaken loose from the plants as they float to the surface of the water and are scooped up by large machinery.

Grapes

One of the fruits that has most inspired man, yet equally often led to this downfall, is the grape. Both of these effects are, however, from wine, the form in which grapes are most commonly consumed. Grapes are being eaten before discovery of their potential for producing a fermented beverage. The most widely cultivated species of grape is *Vitis vinifera* (Vitaceae), a perennial vine native to the eastern Mediterranean region. Many of the table, or eating, varieties of this species produce poor wine, although some, such as the muscats, produce both an interesting wine and a fine eating grape. There are at least 175 table grape cultivates. Thompson seedless and Perlette are among those developed in the United States.

Seedless grapes are easy to eat thus are particularly popular eating grapes. Like most seeded grapes, seedless grapes are propagated by grafting. The most important native New World grape is *V. lambrusca*, the fox grape. Early settlers in the north-eastern part of the country selected the Concord and Catawba grapes from this species. Concord grapes are now the most important grape in America for juice, jams, and specialty wines. This grape initially became popular because of its robust flavour. Pasteurization or processing by ultrafiltration is necessary to prevent grape juice from spontaneous fermentation. Pasteurization, formerly the most common method of preventing fermentation, altere the taste of delicate juices such as that of *V. vinifera*.

The hardy flavour of Concord grapes was able to survive the effects of pasteurization. Consequently, it became the traditional juice grape. In Europe, where ultrafiltration is standardly employed to process grape juice, regular wine grape varieties are generally used. Carefully dried grapes are called raisins. Drying is sometimes carried out in a sulfur dioxide atmosphere to prevent fungal and bacterial growth. Varieties of grapes having a soft texture, reduced stickiness, and pleasing flavour are used produce raisins. Common grape varieties that produce good raisins are the Sultana, Black Corinth, and Muscat of Alexandria. In the United States, almost all raisins are produced is California.

Olives

Although they are mostly considered as martini decorations, olives are an extremely important fruit and source of oil for western people for over 5,000 years. Olives, *Olea europaea* (Oleaceae), are native to the region bordering the Mediterranean Sea where they still grow wild. Olive seeds, dated to 3000 B.C. have been found in excavated settlements in various parts of Israel. Extreme bitterness of fresh olive is due the presence of oleuropein compound that must be broken down or altered to make olives palatable. In ancient times, olives were processed by drying, salting, or pickling. These procedures removed some, but not all, of the bitterness.

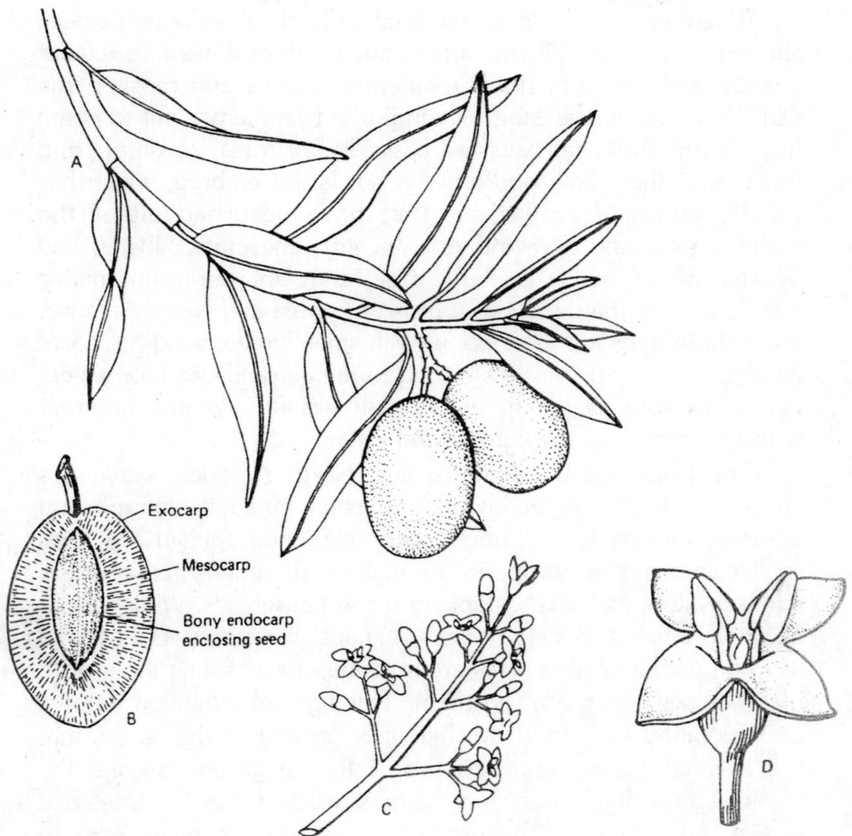

Fig. 6.7. Olives are drupes with a single seed borne on law, stout trees native to the Mediterranean region. A—Branch, B—cross section of a fruit, C—a cluster of flowers, D—enlargement of a fruit.

About 1900, a process was developed in California (where olives had been cultivated since their introduction in 1769) that produced the first truly tender, nonbitter olives.

In this ripe olives are treated with sodium hydroxide and subsequently allowed to oxidize by exposing them to the air, they turn into black olives. Green olives are kept submerged after processing so hydrolysis. Pitting is the last step in processing before the olives are packed in jars or cans. Despite these newer methods of producing edible olives, only about 1 or 2 percent of the world's crop is eaten as fruit. Most olives are pressed for oil.

Temperate Nuts

When we think of nuts, we tend to think of walnuts, pecans, almonds, chestnuts, filberts, and peanuts. All of these nuts except peanuts are native to north temperate regions and most to the Old World. Pecans are the only important temperate nut to come from North America. Nuts are basically dry fruits, so differ from fruits. Also the edible portion of a nut is the embryo, which has greatly enlarged cotyledons that have absorbed all of the endosperm during development. Walnuts, chestnuts, filberts, and pecans are all borne on large trees that are important native components of the deciduous forests of Eurasia and North America. All of these trees are monoecious with small, green, wind-pollinated flowers. Pecans and walnuts belong to the Juglandaceae, or walnut family. Despite its name, the English walnut, *Juglans regia*, is native to southeastern Europe and Asia.

The black walnut, *J. nigra* is a North America native. It's strong taste and exceedingly hard shell make it unimportant commercial species. Walnuts were blanched, pulverized, and soaked in water in Europe in the eighteenth century to provide a milk substitute that was a staple in many households. While walnut fruits were being used to nourish humans, the trees were used to ward off insects. Walnut trees are notorious for inhibiting the growth of other plants surrounding them. This type of inhibition, known as allelopathy, is caused by chemicals present in the leaves that are leached out by rain and soak into the ground around the trunk. Apparently, due to this farmers believed that the trees had insecticidal properties because they planted them near barns to keep flies away from the animals.

But the effect of walnuts on other plants does not extend to insects. In the United States, production of the introduced English walnut is centered in California, but there are also commercial orchards in the northeastern part of the country and adjacent Canada. Pecans and their relatives hickory nuts and butternuts belong to the same family as walnuts but are placed in the genus *Carya*. The most important commercial species is *Carya pecan*, native to Mexico and the American southwest.

Pecans are now extensively grown in the southeastern part of the United States, where they have become a prominent part of the regional cuisine. Thin shells for easy cracking of pecans has developed s a result of commercial breeding program with pecans.

Selection has also improved disease resistance and uniformity of fruit production. Despite its relatively recent introduction into trade, pecans have over 300 named varieties. Annual production in the United States exceeds 52,000 tones. In addition to the pecan, chestnuts were also produced once in a significant quantity in North America. Prior to 1900, roasted American chestnuts (*Castanea dentata*, Fagaceae) were common in the northeastern United States About 1890, the chestnut blight, caused by a fungus of the genus *Endothia*, was introduced into the United States from Asia. The blight virtually destroyed all of the U.S. fruit-bearing trees within a few years.

Besides this American chestnut, there are two other commercially important species, the European chestnut (*C. sativa*) and the Japanese species (*C. crenata*). The Japanese chestnut is

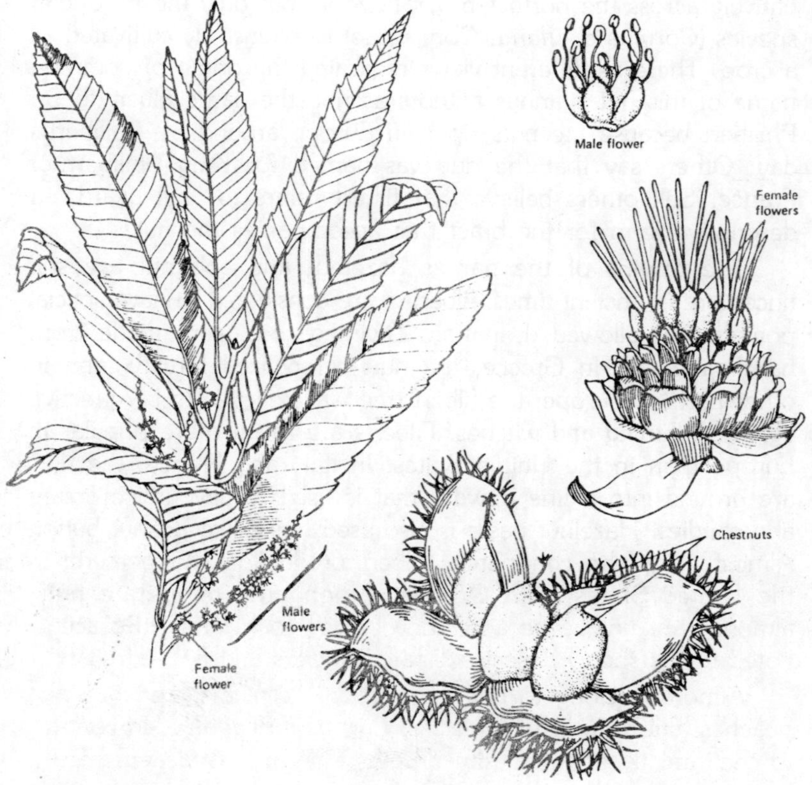

Fig. 6.8. Chestnuts are borne on tall forest trees with coarsely toothed leaves.

more resistant to the fungus than either the American or the European species, is now being planted in North America. Three chestnuts are produced in a cluster. There is a compact influorescence surrounded by fleshy bract with three flowers grouped together, however a single flower produces a nut. When the nuts are mature, the three bracts spread apart, revealing the chestnuts. The nuts are usually picked before they are completely mature, but when the bracts are beginning to separate. Ripening is completed after harvesting.

Unlike most nuts, chestnuts are usually eaten without appreciable drying. In Europe they are considered one of the most versatile and delicious nuts and are used in candies and pastries, are pureed as a vegetable, and eaten as a plain or roasted nut. Hazelnuts, also known as filberts or cob nuts, occur natively across the north temperate zone, but only the European species (*Corylus avellana*, Corylaceae) is extensively cultivated as a crop. There as different views regarding the origin of common name of this nut. Various authors ascribe the name filbert to St. Philibert because the nuts ripen in Europe around St. Philibert's day. Others say that the nut was named for King Philibert of France. Still others believe that filbert refers to "full beard," a descriptive term for the bract that accompanies the nut.

The origins of the names hazelnut and cob nut are still uncertain. In ancient times, filberts were considered to have special powers that allowed divination. Divining rods were made from hazelnut wood. In Greece, the nuts symbolized fertility, and in other parts of Europe, the filbert tree was believed to be effective against lightning and witches. Filberts are much more popular in Europe than in the United States. In Italy and France, the nuts are ground into a fine powder that is used to flavour ice cream and candies. Hazelnut paste is also used much like peanut butter. Almost all of the commercial filbert production (95 percent) in the United States is in Oregon. Among the temperate nuts, almonds are unique because they belong to a family (Rosaceae) noted for its showy, insect-pollinated flowers and fleshy fruits.

Almonds belong to the same genus as plums, apricots, and peaches, but to a distinct species, *Prunus amygdalus*. In contrast to their relatives with fleshy drupes, almonds have a mesocarp that expands little as the fruit matures. Removal of the leathery mesocarp leaves the familiar almond inside the *endocarp (shell)*.

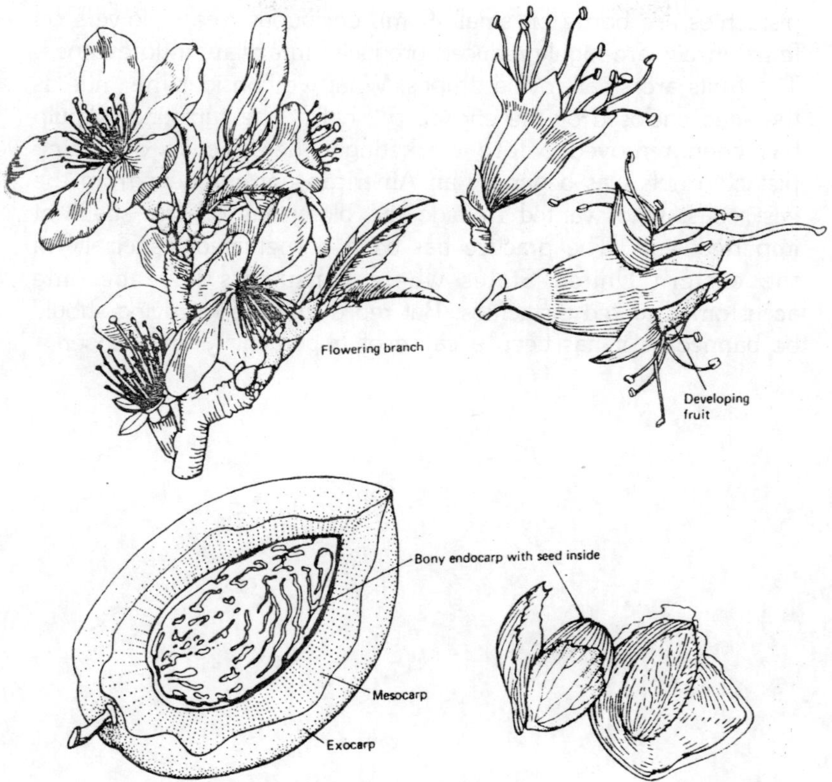

Fig. 6.9. The almond has the same kind of lovely flowers as a peach of cherry tree, but the mesocarp of the almond fruit is tough and leathery.

The characteristic "almond" flavour and odor is due to the presence of a cyanogenic compound, amygdalin in almond seeds, as in that of apricot and peach. The seeds of peaches and apricots are not eaten because they are too bitter for most people's taste and contain dangerously high levels of amygdalin. Almonds are native to Eurasia where they have been cultivated for thousands of years. In the United States, almonds are cultivated in California where the climate is similar to that of the almond's native region. Slightly more exotic than the common dessert nuts is the pistachio.

The distinctive, slightly resinous flavour of the pistachio (*Pistacia veria*), Anacardiaceae, has made it popular in many who love the yellow-green kernels salted or incorporated in ice cream, cakes, or nougat candies. Native to the eastern Mediterranean and central Asia where they have bee cultivated for over 3,000 years, pistachios are borne on small (9 m), deciduous trees. Flowers on female trees are highly reduced produced in axillary inflorescence. The fruits are small, ovoid drupes. What we consider the "nut" is the seed enclosed by the endocarp "shell." The surrounding pulp has been removed before marketing. The practice of dyeing pistachio nuts was begun by an American entrepreneur from the east coast who wanted to hide the blemishes on the shells of imperfect nuts. The practice has been perpetuated, especially in the eastern United States where consumers have become accustomed to red pistachios. But red dyes used for dying should be banned as it has been a cause of hyperactivity and cancer.

7

MEDICINAL PLANTS

Medicinal plants were known to the early civilization. As a matter of fact the history of the durg plants is as old as the hisotory of these civilizations. The Chines have used drug plants quite earlier in 5,000 to 4,000 B.C. The Assyrians, Babylonians, Herbrews and Egyptians knew many drug plants in about 1600 B.C. The works of Greeks, viz., Aristotle (384-322 B.C.), Hippocrates (460-370 B.C.), Pythagoras and Theophrastus (370-287 B.C.) have numerous references of many of the present day drugs. In 77 B.C., a Roma physician Dioscordies wrote 'De Materia Medicia' which described the nature and properties of all the 500 medicinal plants known at that time. His book was was considered to be the most authentic work on medicinal plants for the next 16 centuries or so. There was no advancement in the knowledge of durg plants during the Dark Ages.

After the introduction of printing in Europe in the fifteenth century many persons published 'herbalts' which contained many true and and false informations. Some people advanced the 'Doctrine of Signs of Signatures' which meant that the appearance of a plant or its organs indicated its utility, the sign being placed there by the Creator. Superstition about one such plant Mandrake (*Mandragora officinarum*) to be useful in treatment of human disease as due to its human body like appearance. In the present times medicinal science has paid great attention to the study of drug plants. The branch of medical science which deals with the drug plant is called *pharmoacognosy*, whereas the study of the action of drugs is called *pharmacology*. Usually the morphology of drug yielding plant is the main basis of drugs' classification.

Fig. 7.1. Glycosides ingested by the caterpillar of this butterfly while feeding on milkweed are stored in the body and, after metamorphosis, appear in the adult monarch's body.

Drugs Obtained from Roots

Rauwolfia

The drug Rauwolfia is obtained from the roots of *Rauwolfia serpentina* of *Apocynaceae*. It is native of India. The genus is found growing in the tropical regions of Asia, Africa and America. The Asian countries are India, Bangla Desh, Sri Lanka, Burma, Malaysia, Thailand and Indonesia. Five species of *Rauwolfia* grow in India. *R. Serpentina*, the most important of the five species is found in the sub-Himalayan tract from Punjab through Nepal, Sikkim, Bhutan to Assam, eastern and western ghats, Central India and the Andamans. The plant is grown commercially in Uttar Pradesh, Bihar, Orissa, West Bengal, Assam, Andhra Pradesh, Tamil Nadu, Karnataka, Keralan and Maharashtra. The plant is an erect perennial shrub. It is of a height of ½ ft. to 1½ ft. and may sometimes attain a height of 3 ft. The plant is evergreen. Leaves are in whorls of three. Flowers are small white or pink and are borne in cyme in large number. The fruit is a drupe.

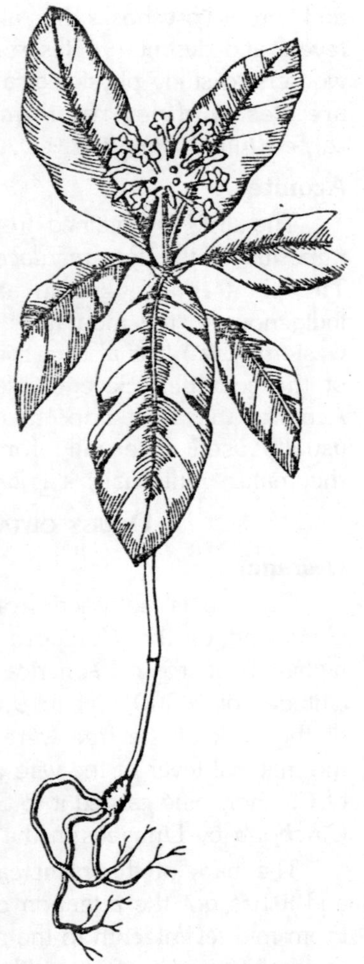

Fig. 7.2. The snakelike root of Rauwolfia contains an alkaloid used in the treatment of hypertension and schizophrenia.

Hot and humid climates of the tropics is best suited for its growth. It prefers shady habitats of the forests. It grows well in areas of very high rainfall (250 cm to 500 cm) and temperature range of 10 to 38°C. It can grow in different kinds of soils ranging from sandy alluvial loam to red lateritic loam. Well drained humus rich clayey soil is best for its growth. The plant is usually propagated by means of seeds. Rauwolfia drug is obtained from the root of the plant. The bark of the roots yields several alkaloids (about 80), the most important of which is *reserpine*.

Reserpine is extracted from another two species—*R. vomitoria* of Africa and *R. tetraphylla* of America. Reserpine is of great medicinal use in treatment of violent kinds of insanity and high blood pressure. It acts as a sedative and depressant in hypertension and chronic psychoses. It is also used for treatment of insect bites, fevers and dysentery. Reserpine is now widely used by research workers to study physiological systems. Other important alkaloids are *deserpidine*, *rescinnamine*, *reserpinine*, *serpentine*, *serpentinine*, *ajmaline*, etc.

Aconite

The drug is obtained from the tuberous roots of *Aconitum napellus* of the *Ranunculaceae*. The plant is a perennial herb. The plant is also known as monk's hood or wolfbane. It is indigenous to the mountains of Europe (Alps, Pyrenees, etc.and western Asia.) It is also cultivated as an ornamental plant is most of the countries. Several alkaloids are obtained from the roots. Aconite, the most important alkaloid, is poisonous. Aconite is usually used externally for the treatment of neuralgia and rheumatism. Internally it is taken to relieve pain and fever.

Drugs obtained from Barks

Quinine

Quinine is obtained from the bark of several species of *Cinchona* of the *Rubiaceae*. The plant is native to Andean highlands of tropical America. It still grows in Peru and Bolivia at altitudes of 3,000 feet to 9,000 feet. The medicinal properties of the bark of the tree were discovered in 1638 when it cured the malarial fever of the wife of the Viceroy of Peru, the Countess of Cinchon. She carried it to Spain in 1639. The plant was named *Cinchona* by Linnaeus in the eighteenth century after her name.

The bark of the plant called Jesuit's bark or Peruvian bark and it has got the attention of the people throughout the world. From mid-seventeenth to the mid-nineteenth century the wild trees in south America were ruthlessly cut down to get the bark. The British and the Dutch secured the seeds from South America and started large plantations in India and Java respectively. The Britishers had collected seeds from *Cinchona succirubra*. The Dutch plantations of Java were established from the seeds of *Cinchona ledgeriana* named after a British resident of Bolivia, Charles Ledger, who had collected the seeds. About 90 percent of the world export trade in quinine is carried on by Java. Other species of Cinchona used for the drug are *C. officinalis* and *C. calisaya*. The tea plantation has replaced the plant in Sri Lanka because the drug has quite uneconomic price.

Fig. 7.3. A flowering and fruiting branch of Cinchona officinalis, the of which yields quinine, a remedy for malaria.

The other *Cinchona*-growing regions are Burma and Tanganyika (Tanzania). *Cincona* plant is a fairly large tree and may attain a height of fifty feet or more. Depending upon the species of plant, the bark covering the stem can be light coloured, pale, brown or dark brown. The leaves are opposite. Yellow or pink coloured flowers are borne in terminal panicles. They are produed from the third or fourth year of growth. Cross pollination takes place. The plants are limited in distribution. They grow in the topics between 10° north and 20° south of the Equator at altitudes more than 1,000 feet. They grow well in regions of very high temperature and high rainfall (60 inches or more). Good porous soils are good for their growth.

Cinchona is propagated by seeds. Mixing of characters and consequent variation in plant is attained as a result of natural hybridization. It is now usually propagated by means of graftings and cuttings. The trees are usually felled down after they are 10 years old since the bark develops maximum alkaloid content by that time. The bark is removed from the stems, branches and roots and is slowly dried at not very high temperature. The dried bark is sent to the factory where the alkaloids are extracted by solvent extraction.

Cinchona bark has four important alkaloids (totaquine) which are quinine, quinidine, cinchonine and cinchonidine. Quinine is the most important of the alkaloides. *C. ledgeriana* has the maximum percentage of quinine in relation to the other alkaloids. Quinine is a antimalarial drug and is very bitter and white granular in appearance. It is also useful as a tonic and antiseptic. Some synthetic antimalarial drugs are also manufactured these days.

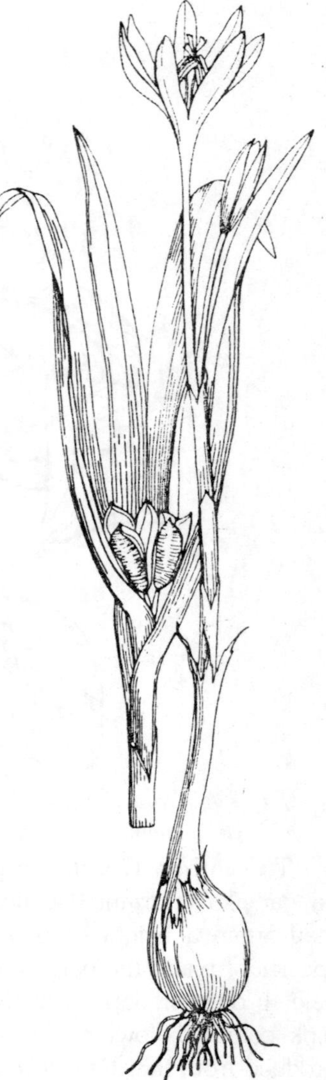

Fig. 7.4. The autumn crocus yields colchicine used to treat gout and cell malignancies.

Drugs obtained from Stems

Ephedrine

The drug ephedrine is an alkaloid. It is obtained from several species of *Ephedra* of the *Gnetaceae*. Two important species are *E. sinica* and *E. equisetina*. The plant is native of Asia. In China it has been in use for over 5,000 years. The plant is leafless shrub with green stems. The plant is woody and xerophytic and grows in arid regions of the world. The plant is dioecious. Entire plant is the source of the alkoloid. However, its importance as a source of ephedrine has been reduced due to synthetic manufacture of the drug. The drug is used in the treatment of nasal and bronchial congestion, colds, asthma, hay fever and other ailments. It also acts as a stimulant.

Drugs obtained from Leaves

Eucalyptus

Eucalyptus oil is used in medicine. It is obtained from the leaves of several species of *Eucalyptus*. The plant is native to Australia. It is also grown in the Mediterranean region. U. S. A. and many other places. The plant is a tall tree which attains a height of 200 to 300 feet. The leaves are scythe-shaped in *E. globulus* and of different shapes in other species. Eucalyptus oil is used in the treatment of cold, malaria, nose and throat troubles and fevers.

Cocaine

Cocaine obtained from the leaves of a South American plant, *Erythroxylon coca* of the *Erythroxyllaceae*. The plant is indigenous

Fig. 7.5. A branch and flower of a coca plant, which yields the alkaloid cocaine.

to the Andean regions of Peru and Bolivia. The plant is now widely cultivated in the tropical regions of South America, Java, Sri Lanka and Taiwan (Formosa). The plant grows are higher altitudes. The plant is a small, much branched shrub or tree. Small, sessile, elliptical leaves are alternately arranged. The flowers are borne is small axillary groups. The leaves are harvested by hand when the trees are two to three years old. The leaves are picked three or four times in a year. Dried leaves are immediately exported in boxes. The alkaloid cocaine ($C_{17}H_{21}O_4N$) is ususlly extracted in the importing countries by solvent process. The drug is chiefly used as a local anesthetic. It is also used as a tonic for the digestive and nervous symptoms. The leaves are chewed by the natives of South America for stimulating physical and mental activities. The coca is 'de-alkaoidized' in U. S. A. to be used as 'cola' flavourings.

Digitalis

Digitalis is obtained from the dried leaves of foxglove (*Digitalis purpurea*) of the *Scrophulairceae*. The plant is a native of Southern and Central Europe, which continues to be its chief grower. However it is grown as beautiful ornamental plant in other parts of world. The plant is a pubescent biennial or pernnial herb. The plant stem can attain a height of 5 feet but leaves are crowded in lower portion of stem. The stem bears purplish flowers in a terminal spike. The leaves are dried in shaded conditions. The active glycosides are extracted fron the dried leaves by solvent (alcohol) process. The glycosides are digitoxin, digitaline and digitalein and digiton. They are very important as stimulants of heart and as diuretics. Digitalis tone up the circulatory system by inducing the heart to make powerful and complete contraction.

Belladona

The drug is obtained mainly form the dried leaves of the deadly nightshade, *Atropa belladona* of the *Solanaceae*. The plant is indigenous to Europe and Asia Minor. It is widely grown in Europe, U.S.A. and India. The plant is a perennial herb with a creeping root stock. It bears alternately arranged ovate leaves. The stem is hollow. The flowers are purplish. The plant bears brownish or blackish berry. The leaves are harvested at the flowering time (May-June). The leaves are then dried at low temperature for a period of 2 to 15 weeks. The alkaloids present in the plant are extracted with the help of solvents.

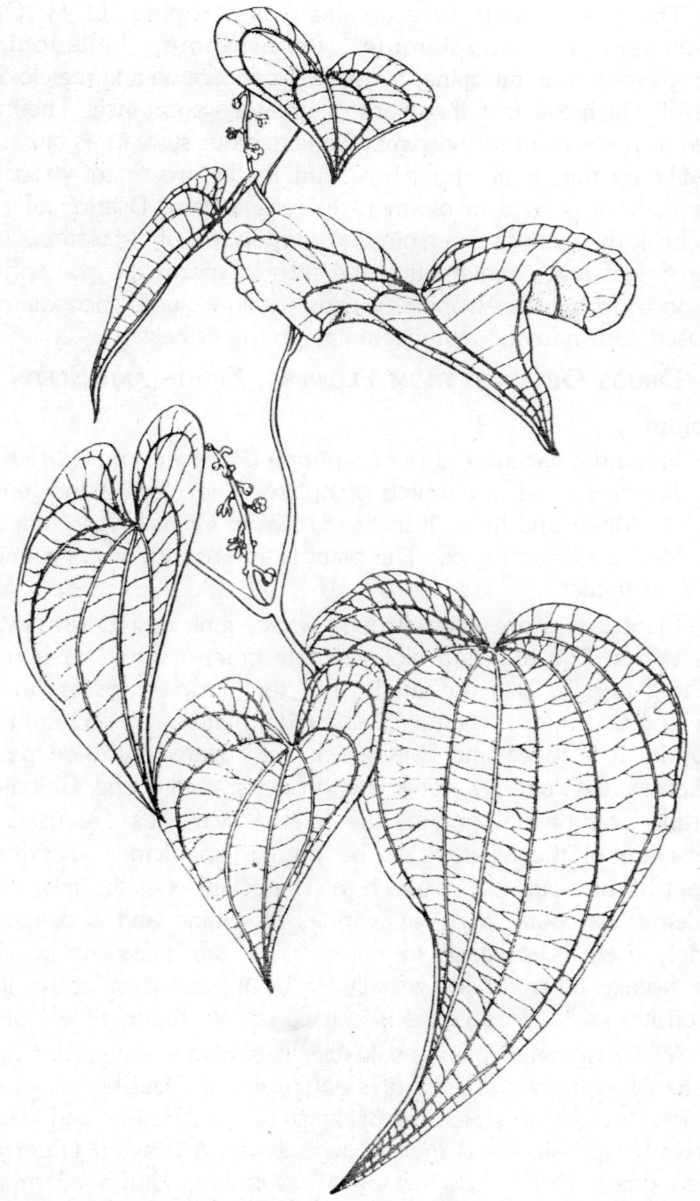

Fig. 7.6. The roots of yam vines are the principal sources of steroid precursors used to produce active compounds in oral contraceptives, and to treat hormone imbalances and heart.

The plant yields several alkaloids–atropine ($C_{17}H_{22}O_3N$) hyoscyamine, scopolamine, apoatropine, belladonine, norhyocyamine noratropine, hyoscine, tropacocaine and meteloidine out of which the first three are of greater importance. They are used as a stimulant to the sympathetic nervous system, as diuretics, in dilating the pupil of the eye and in the treatment of palsy. Externally it is used in ointment to relieve pain. Dilation of eye, to check the excessive perspiration, stimulation of circulation, local pain releaf and counteracting the muscle sperm are the various purposes for which Atropine or hyoscyamine is used. Scopolamine is used as a narcotic anti-insomniac and anesthesia.

Drugs Obtained from Flowers, Fruits and Seeds

Opium

Immature capsules of opium poppu (*Papaver somniforvm*) of the *Papaveraceae* are source of opium. The plant is indigenous to Asia Minor and India. It is also grown in China, West Asia and the Mediterranean region. The plant is an annual herb of a height of 2 to 4 feet.

Plant leaves are alternate and white, pink or reddish flowers are borne. The fruit is a globular pale green capsule. A number of fine parallel slits are made into the unripe capsules in the evening and in dry weather. The exuding latex is collected in the moring. It is rolled into balls which are covered by dried petals. Whereas the 'alba' varietes are grown in India and China for getting commercial opium, the 'glabra' varieties are used for exraction of the alkaloids to be used in medicines. There are about 30 alkaloids out of which the important ones are morphine, codeine, narcotine and papaverine. Morphine and codeine are widely used as sedatives to relieve pain and cause sleep. They are usually taken orally, rectally or by injections to cause local insensitiveness. Morphine is also used in the treatment of cough.

While opium is a very valuable, it is also a dangerous drug. Opium has narotic effects. It is eaten and smoked by millions of people for deriving pleasant feelings of exhilaration and peace. However, physical and mental degradation debility, delirium and even death can be caused as a result of addition of opium, morphine codeine, heroine (artificial derivative of morphine) etc. The drug should be used in very limited quantities and under strict supervision of physician.

Fig. 7.7. The opium poppy is the source of morphine, papaverine, and codeine.

Strychnine

The drug is obtained from the seeds of *Strychnos Nuxvomica* of the *Loganiaceae*. The plant is indigenous to Australia, India, Sri Lanka and Cochin China. The plant also grows widely in South America and Europe. It was introduced in Europe in the 16th century. The plant is a woody vine of the tropical forests. The stem bears ovate leaves in opposite manner. The plant bears small flowers and large indehiscent fruits. A fruit of *Strychnos* contains 3 to 5 hard and greyish seeds.

The alkaloids present in the seeds are strychnine ($C_{21}H_{22}N_2O_2$) and brucine ($C_{23}H_{26}\ N_2O_4$). They are expressed from the seed with boiling sulphuric acid. The alkaloids are first precipitated and are then purified. *Strychnine* is a very poisonous substance. It is of medicinal uses in the treatment of nervous disorders and paralysis It stimulates the central nervous system and acts as a tonic. It is famous as arrow poison (curare).

Fig. 7.8. The leaves of the garden ornamental foxglove which yields digitoxin, a steroidal glycoside effective in stabilizing the heart's action.

Drugs Obtained from the Flower Plants

Antibiotics

Antibiotics are produced by molds, actinomycetes and bacteria. They kill pathogenic bacteria. Some of the antibiotics produced by such micro-organisms are as follows :

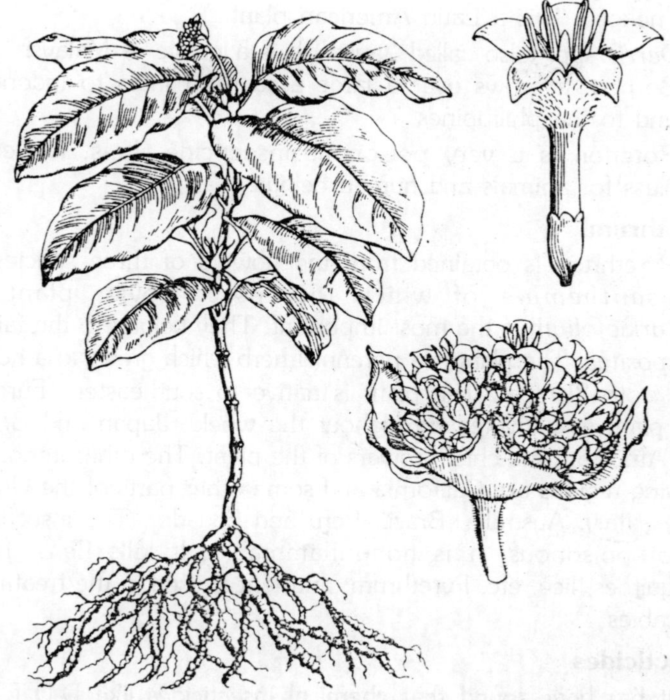

Fig. 7.9. Cephaelis, from which ipecac is obtained.

Penicillin is obtained from *Pencillium notatum*.

Streptomycin is obtained by *Streptomyces griseus*.

Aureomycin is produced by *Streptomyces aureofaciens*.

Chloromiycetin is obtained from *Streptomyees venezuelae*.

Terramycin is obtained from *Streptomyces rimosus*.

Neomycin is obtained from *Streptomyces fradiae*.

OTHER DRUGS

Kelp comprise certain brown algae, viz., *Macrocysti pyrifera*, *Laminaria digitata*, *L. saccharina*, etc. They yield iodine, potash, acetone, sodium alginate, etc. Fruiting body of a fungus, *Claviceps purpurea* gives Ergot. *Agar* is obtained form red algae, e.g. *Gelidium corncum*, *Euch euma spinosum*, *Gracilaria lichenoides* and some other species.

Rotenon

Rotenone is a poisonous substance widely used as insecticide and is obtained from roots of two leguminous limbers. The plants are:

(i) *Lonchocarpus nicou* (also called 'cube', 'timbo', 'haiari', and ('nekoe'). It is a Latin American plant.

(ii) *Derris* spp. (also called 'tauba') It is a native of Malaysia and Borneo. It grows over a large area from India to Indonesia and to the philippines.

Rotenon is a very poisonous insecticide. It is, however, harmless for animals and human beings.

Pyrethrum

Pyrethrum is obtained from the flowers of three species of *Chrysanthemum* of which the insect flower plant *C. cinerariaefolium* is the most important. They belong to the family *Compositae*. The plant is a perennial herb which grows to a height of 1½ to 2½ feet. The plant is native to southeastern Europe. The plant is cultivated throughout the woeld. Japan and British East Africa are the chief growers of the plant. The other important growing regions are California and some other parts of the United States, Italy, Australia, Brazil, Peru and Equador. The insecticide is not poisonous. It is non-inflammable. It kills fleas, flies, mosquitoes, lice, etc. Pyrethrum ointment is used in the treatment of scabies.

Insecticides

It has been found that chemical insecticides like D.D.T. are also toxic to the human beings, search for non-toxic botanical insecticides has been going on all the time. Over 1200 species of plants have been found to be insecticidal. Pyrethrum and rotenon are two important insecticides obtained from plants.

8

Tanning and Dye Materials

Tannins

Tannins are complex organic compounds which are astringent and acidic. They are glucosidal in nature and consist of carbon, hydrogen and oxygen like the carbohydrates. They are water soluble. Their exact role in plants is not understood. They are, however, of great economic importance. They convert animal hides into leather by initing with their proteins. The leather thus formed is stable, resistant to air, moisture, temperature and bacterial action. The basis of ink, dark blue or greenish-black compound is also obtained from tannins by it reaction with iron salt. Tannins find use in medicines because of their astringent properties. They are also used in oil drilling to reduce the viscosity of the drill.

Tanning Industry

Tanning of hides was done about a thousand years before Christ. The tanning of hides is done in a slow and gradual manner so that tannin reaches every part of it. The hide is first dipped in used tannin liquors, an increased concentration of tannin enters into it as the hide swells due to the acidity of tannin liquors. The entry of tannin to the inner part is, however, not possible if the hide is dipped in a strong solution of tannin because the latter causes the leather to become impervious on the other surface. The leather after tanning, is treated with oil or grease, sye, and some finishing materials like gum, resin, albumen, etc.

Sources of Tanning Materials

Tanning is found in almost all the plants but only some species

Fig. 8.1. A branch of an annatto plant with flowers and fruit.

contain substantial amounts. Tannins can be obtained from the following sources :

1. *Leaves*. Sumac (*Rhus* spp.), Sicilian dumac (*Rhus coriatia*) and Gambier (*Uncaria gambir*).
2. *Fruits*. Myrobalan nuts (*Terminalia* spp.), Divi-Divi (*Caesalpinia coviaria*), tara (*Caesalpinia spinosa*), algarobilla (*Caesalpinia brevufolia*), and Turkish oak (*Quercus macrolepis*).
3. *Roots*. Tanners dock (*Rumex hymenosepalus*), and plametto (*Sabal palmetto*).
4. *Bark*. Hemlock (*Tsuga canadensis*), oak (*Quercus* spp.), mangroves particularly red mangrove (*Rhizophora mangle*), water (*Acacia* spp.) larch (*Larix decidua*) avaram bark (*Cassia auriculata* and babul (*Acacia nilotica*).

5. *Woods*. Chestnut (*Caslanea dentata*) and Quebracho (*Schinopsis* spp.)

Manufacture of Ink

Ink is another item obtained from tannin and bears a commercial importance. The has been in use since 2600 B. C. in China. Egyptians are also known to have used ink for writing on papyrus in 2500 B.C. Of the several types of modern inks, carbon and tannin ink are most important.

Coloured inks

Coloured inks are obtained by mixing Aniline dyes or natural dyestuffs with water, alum and a gum.

Tannin inks

Tannin combines with iron salts to give a blue-black colour. Tannin derived from the insect, galls formed on the twigs of Alepppo oak (*Quercus infectoria*) is the chief source of tannin inke.

Carbon inks

Chinese or *India ink* and *printing ink* are examples of carbon inks. They remain on the surface of paper like paints. Soot, carbon black or lampblack obtained by burning a vegetable oil or pine wood and glue, gum arabic etc. are the sources of China ink. Carbon for the printing ink is prepared from petroleum, natural gas, etc. and is mixed with a drying oil (linsed oil), rosin, some chemical drier and soap.

Dyes and Pigments

Vegetable dyes have been in use since times immemorial. Egyptian tombs had colouring of saffron, safflower, indigo and madder. The primitive people of Britain decorated their skins with a blue dye. After the discovery of an indogo-type dye from 'useless' coal tar component by *Perkin* (1856), vegetables oils are now useless. The synthetic or aniline dyes obtained from coal-tar products are brighter, more permanent, cheapter, easier to use and afford a wide range of colours. An account of the vegetable dyes is, therefore, of only historical interest. Only a few natural dyes have been able to compete with the synthetic ones.

Dyes are chiefly used in textile industry for giving fast colours to the fabrics, and for colouring paints, varnishes, leather ink, paper wood, furs, food, cosmetics medicines and toothpastes. The dyes are treated with mordants, salts of metals like aluminium, iron, chromium, etc. for making them insoluble. The mordant forms a fine layer of metallic oxide on the fabric. The oxide along with the dye from an insoluble compound which helps the dye to remain fast on the cloth.

Sources of Natural Dyes

Natural dyes are obtained from various parts of a number of plants. The sources are as follows :

WOODS

Acacia

The plant belongs to the *Leguminosae* family. *A catechu* yields a dye called 'Katha'. By boiling the heartwood of the plant

Fig. 8.2. Wood, a source of a blue dye, was one of the dyes used to make the green outfits.

with water. the dye is obtained. The dye is obtained from the heartwood of the plant when it is boiled with water. It is largely used in printing and in colouring pulp and paper. It is also used for dyeing cloth. It is often used with diazo salts to give bright shades. Dyeing the ship sails with 'katha' protect them from weather and sea water.

Pterocarpus

The plant belongs to *Leguminosae* family. *P. santalin* yields a red dye. *sanitalin*. The red dye is obtained from the wood and root of the plant. The dye is used to colour idols etc. Natives of Ododobo, Africa used to paint their bodies with the dye. The dye is dissolved in alcohol to give a salmon pink colour to the fabrics. The dye is also obtained from *P. tinctorius*. It was used to colour the newly born children by the natives of Pungo Andongo, Africa. The native women also used the dye for colouring their feet to give the appearance of shoes or slippers.

Haematoxylon campechianum (Logwood)

The dye is knonw as *haematoxylon*. It is the product of heartwood of the leguminous tree *H. campechianum*. It is obtained from the heartwood of the leguminous tree, *H. campechianum*. The dye is about 10 per cent in content. It is violet, grey, blue cr black in shade. As the dye is capable of staining nuclei it is widely used in histological studies. It is also used in dyeing silk, wool and cotton.

LEAVES

Isatis

I. tinctoria yields a blue dye called *wood*. The dye was very popular in Europe before it was replaced by the indigo. The dye was called *universal dye* during the Middle Ages. The cloth used to be dyed blue with this dye and yellow with a dye obtained from *Reseda luteola*. Such a dyed cloth was called 'Saxon green'. The plant is a native of Tibet and Afghanistan. The leaves of the plant are crushed into a pulp which is then made into balls. The balls are dried and fermented to yield the colouring principle the indican.

Indigofera

The plant belongs to the *Leguminosae* family. It yields the dye, *indigo* ("Neel"). It is called the "king of dyestuffs". The dye has been known to the peope of Asia for over 4000 years. The word indigo is derived from a Latin word indicum. The dye is known to be very much in use in Persia in 600 A.D. Arabs brought the Indigo to Mediterranean regions to carry on the trade in Indigo. The dye became very popular in Europe. The dye reached China from Persia. India was a leading producer of indigo during the reign of Eat India Company and later British Government. During the freedom movement for the liberation of our country from British rule Mahatma Gandhi championed the cause of indigo cultivators of the Champaran district of Bihar against their exploitation by Britishers.

Indigofera tinctoria is the principal source of the dye. The plant is a native of southern Asia. *I. arreca* is another important source of the dye. A colourless glucoside, called Indican is present in plant leaves. During fermentation of the leaves indican is hydrolysed to glucose and indoxyl. Oxidation of indoxyl results in Indigotin formation. If the dye is high quality it is deep violet-blue and porous. The nature dye has now been completely replaced by the synthetic one. The dye is used in dyeing and printing cotton fabrics, rayon and wool. It is used in washing of cotton clothes. Pigments for printing inks, lacquers and plants are also manufactured from it.

Lawsonia

It belongs to the family *Lythraceae*. A red coloured dye is obtained from the leaves of *Lawsonia alba* ('Mehndi'). The plant has been known to Indians from the remote past. The dye is favourite with Indian women and is used for colouring palm, feet, finger with it on festive occasions. The leaves are crushed in the form of a paste for this purpose. The hair is also dyed with it.

Roots and Tubers

Morinda

The plant belongs to *Rubiaceae*. A yellow dye is obtained from the barks of the roots of *M. citrifolia* and *M. angustifolia*. The plant yield optimum amount of dye when the plants are three or four years old. The dye protects paper, cloth, etc. form white ants.

Rubia

A red dye is obtained from the plant. *R tinctorium* and *R. cordifolia* are the two important species of *Rubia*. *R. tinctorium* (Madder) was known to ancient Indians, Persians and Egyptian. The clothes of the Egyptian mummies were dyed wiht the dye.

Libyan women dyed their dresses and Alexander the uniform of his soldiers with it. The dye of Rubia used in our country is obtained from *R. cordifolia* growing in Himalayan region. It is used for dying fabrics.

Curcuma

The turmeric plant belongs to *Zingiberaceae*. The tuber of *C. longa* is the source of an orange red dye. The colouring matter is known as *curcumin*. The dye is very important in the tropical countries. It is widely used in our country for colouring food particularly pulses, vegetables, etc. It is used in religious functions of the Hindus and is believed to be very auspicious.

FLOWERS

Carthamus

Safflowr (*Carthamus*) Plant belongs to the *Compositae* family. The plant finds reference in Ssnskrit works as 'Kusum' or 'Kusumbha'. The florets of *C. tinctorius* provide a scarlet red dye. The Egyptians dyed the grave cloth for the mummies with this dye. The dye is used in our country to dye cloth for getting cherry red colour. The dye is also used for imparting colour to biscuits, waffers, cakes, etc.

Crocus

Saffron plant belongs to the *Iridaceae* family. A yellow dye is obtained from the orange coloured stigmas and the upper part of the style of *Crocus sativus*. Saffron was well known to the ancient Greeks and Romans. It also finds reference in Sanskrit literature. Kashmir valley of our country is the only place where the plant is grown. Saffron dye is used for colouring rice and flavouring vegetable dishes. The dye is also used on ceremonial occasions. Being considered to be auspicious, it is applied on forehead.

Other plants. Butea monosperma and *Nyctanthes arbor-tristis*.

FRUITS

Terminalia

The plant belongs ot *Combretaceae* family. Yellow or black dyes are obtained from the fruits of *T. chebula*. A permanent yellow colour is obtained from it in combination with Alum. Treatment of the dye with iron salts in black colour.

Other plant. Rhamnus infectorius.

Seeds

Bixa

A bright yellow or flesh coloured dye is obtained by macerating th epulp (aril) of the seeds of *B. orellana*. It is used for dyeing silk and for colouring butter, cheese, ghee etc.

Gum-Resin

Garcinia

It belongs to *Guttiferae* family. The dye, 'gamboge', is obtained from the bark of *G. hanburyi*. The dye is used in the water colours and gold coloured spirit varnishes for metallic surfaces. The dye is used for dyeing silk cloth. Bark of *G. spicata* is the source of another similar dye Japanese dyestuff.

Lichens

Lichens like Parmellia *omphalodes* etc. yield dyes (orchil, etc.) which are used for dying clothes. Litmus is obtained from *Roccella titictoria*. It is an indicator of the acidity and is used in chemical studies.

9

SUGARS AND STARCHES

Sugars and starches, the two common forms of carbohydrates, are reserve food supply of not only plants but animals too. Both are necessary food for man. They are most valuable products of the plant world. The sugars are soluble but starches are insoluble carbohydrates.While the principle sources of sugars can be stem or underground roots, those of starches are cereal grains and certain underground tuberous root or stem structures. The principle sugar is sucrose. The sources of storage sugars are:

1. Roots—Beets, carrots, parnships etc.
2. Stems—Sugarcane, maize, sugar maple, sorghum, etc.
3. Flowers—Palm.
4. Bulb—Onion.
5. Fruits.

The sources of starches are (i) cereals like maize, wheat, rice, etc. (ii) potato, aroids; tapioca, arrow, root and sago.

SUGARCANE (*SACCHARUM OFFICINARUM*)

Origin and History

Sugarcane and sugarbeet are principal sources of sugar. Dependencey of entire sugar industry on them make them to be called industrial plants. More than half of world's sugar supply is obtained from Sugarcane. The best production of sugarcane is in the tropics. Sugarbeet (*Beta vulgaris*) is cultivated in the cool temperate zone which is not suited for the sugarcane. Sugarcane is believed to have originated in the tropics in or near New Guinea in the pre-Christian times. The plant has not been found in a wild state, and, therefore, its wild ancestor is not known.

In 327 B.C. *Nearchus*, an officer in Alexander's army, wrote that the barbarians beyond the Indus were able without the help of bees to make from honey the sap of a honey-bearing reed. On the basis of chromosome number (2n = 80) and morphology *S. robustum* and *S. officinarum* still occur in plenty in the New Guinea area today. *S. officinarum* ('noble') canes migrated to Asia. There it hybridized with the local *S. spontaneum* to produce *S. sinense*. These hybrid thin canes grow very well in the monsoon climates there. *S. officinarum* also spread to the Pacific Islands. The plant was carried to Egypt in 641 A.D. and to Spain in 755 A.D. The Spanish and Portuguese explorers spread the plant to Madeira in 1420. Cultivation of the crop begun by the end of sixteenth century, today sugarcane is widely cultivated in almost all the moist tropical and semi-tropical regions.

Sugarcane Growing Regions

The chief sugarcane-producing regions lie between 30° north and south of the Equator, viz, India, Cuba, Brazil, Philippines, Hawaii, Java, Puerto Rica, Australia and U.S.A. The first three countries are the major producers of sugarcane. In India, Uttar Pradesh and Bihar are by far the leading states in growing sugarcane. Sugar cultivation area in India is more than 4 million acres of land.

Botanical Description

Sugarcane belongs to the family *Gramineae*. The plant is a coarse grass. It is quite tall. It has a height of 10—20 ft. A tuft of a number of stems may be present in a single plant. The stem is solid and jointed. Arising from the several root initials of the lower nodes, prop or stilt roots support the stem. Axillary buds are present in nodes. The stem bears a terminal inflorescence, but only few seeds are formed. The cultivated varieties of sugarcane are heterozygous and do not breed true. The plants are, threfore, propagated by vegetated by vegetative means.

Ecological Factors

Sugarcane is easy to grow. It needs a hot and moist climate (temperature averaging 70°F and an annual rainfall around 60 inches which may be supplemented by irrigation). Rich, deep and well drained soil gives best yield.

Cultivation

Almost mechanical method is employed for cultivation in modern days. In poor countries, however, cultivation may be done by hand at all stages "Sets" cut from the upper part of old stems

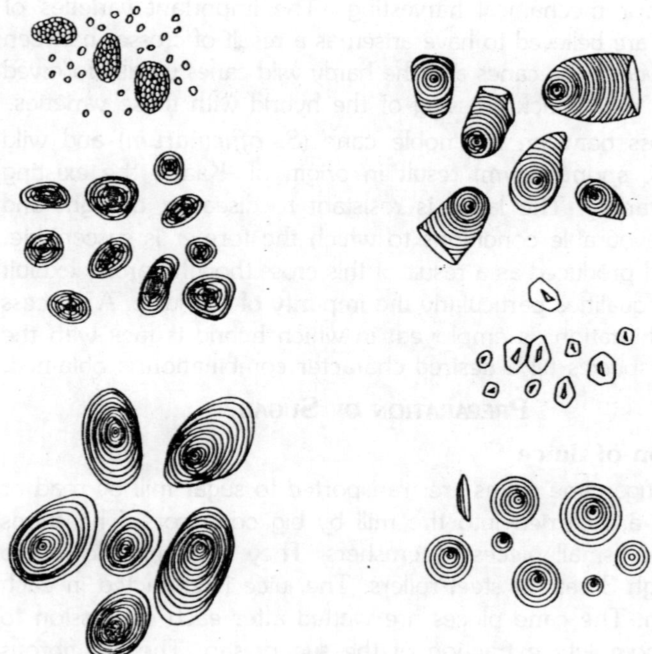

Fig. 9.1. Starch granules of potato, bean, rice, sago palm, corn and wheat.

are usually used for cultivation. A 'set' or 'seed' has several nodes and internodes. The field is plowed deeply and 'sets' are placed in the furrows or trenches and are covered with soil. A sufficient supply of nitrogen, potash and phosphate fertilizers is given to the soil. In about two weeks time the 'sets' sprout, i.e., the axillary buds from the upper nodes give rise to new shoots and the lower nodes produce rings of adventitious roots. Usually 2 or 3 crops are given by 'set'. It has, however, been found to produce aerial shoots (ratoon crops) continuously for as many as 20 years, under exceptionally favourable conditions. Great care is required to promote profuse tillering.

As soon as the tillers have developed their own roots, they are severd from the mother shoot and are carefully planted separately. As sugarcane is not able to compete with weeds, constant weeding is necessary. Fertilizers shoould also be supplid to the field during the early stages. Harvesting is done usually after one to one and a half year of the planting of the 'set'. Harvesting is usually done by hand by chopping the stema at the base with a cane knife. In advanced countries cutting machines are used for mechanical harvesting The important varieties of sugarcane are believed to have arisen as a result of crosses between the soft sweet nobe canes and the hardy wild canes usually followed by one or more back crossing of the hybrid with these varieties.

A cross between the noble cane (*S. officinarum*) and wild species (*S. spontaneum*) result in origin of 'Kaab', the existing superior variety. The latter is resistant to diseases, drought and other unfavourable conditions to which the former is susceptible. The hybrid produced as a result of this cross though hardier exhibit some bad qualities particularly the impurity of the juice. A process called 'nobilization' is employed, in which hybrid is fack with the cultivated species till a desired character combination is obtained.

PREPARATION OF SUGAR

Extraction of Juice

The sugarcane stems are transported to sugar mill by road or rall. They are carried into the mill by big conveyors. The stems are cut into small pieces by crushers. They are then allowed to run through 3 sets of steel rollers. The juice is extracted in each expression. The cane pieces are wetted after each expression to facilitate complete extraction of the sugary sap. The dry fibrous residue is called *bagasse*.

Purification of Juice

Juice is dark-greyish and sweet containing a number of impurities. It contains beside sucrose and glucose, small qualities of protrins, gums, organic, acids, pectin, ash, minerals, dust particles, pigments and small pieces of cane. The soluble non-sugars are precipitated (*clarification*) and the insoluble substances are removed (*defacation*) to purify the juice. After filtering to remove the soild particles the juice is heated to coagulate the proteins. Neutralization of organic acids is performed by adding lime. However it is also added to prevent the sucrose to be converted into some inferior sugars. In addition to lime, chemical compounds like sulphur dioxide, sodium carbonate, carbon dioxide, phosphoric acid, etc., are added to the cane juice to precipitate out the non-sugars. Insoluble particles are removed from partly purified juice by running it through filter presses.

Concentration

The juice is boiled in vaccum pans to concentrate it into a thick syrup ("*massecuite*") from which a solid mass of brown crystals separate out. The liquid remains called *molasses* is separated out through perforated bases or by using centrifuges. The molasses may be further evaporated to form sugar crystals or can be used as a cattle feed or can be fermented into rum, ethyl alcohol and vinegar.

Refining

The brown crystals (96% sucrose) are further refined to give the white crystals of sugar. The brown crystals are dissolved, washed, recrystallized and are decolourized with carbon. The refined sugar is transported usually packed in gunny bags. Sugar is also packed in the form of powder and cubes.

Uses of Sugar

1. The byproduct, molasses, is of great industrial value. It is used in the distallaries for the manufacture of rum, industrial alcohol and vinegar.
2. The molasses is very good cattle feed.
3. The molasses is also sometimes used as a manure.
4. It is the chief sweetening material. It is extensively used all over the world for sweetening beverages, various kinds of food preparations, etc.
5. The molasses used in cooking and candy making.
6. Sugar mills use the bagasse as fuel. It is also a source of paper and as an ingredient of fibres board.

OTHER SOURCES OF SUGARS

Palm Sugar

Several species of palm are a very important source of sugars in the tropics. The sugar is derived from the tender parst of the stem (date palm) and flower of wild date (*Phoenix sylvestris*), the palmyra palm (*Borassus flabellifer*), the coconut palm (*Cocos nucifera*), the toddy palm (*Caryota urens*) and the gomuti palm (*Arcnga pinnata*).

Sorghum Syrup

It is derived from the stem of the sweet sorghum (*Sorghum vulgare* var. *saccharatum*).

Maple Sugar

The maple-sugar is derived from the sweet sap of the stems of sugar male (*Acer saccharum*) and the black maple (*A. nigrum*).

10

Industrial Plants: Rubber and other Latex Products

Rubber

Ruber is indigenous to South America. Rubber has a very interesting history. The rubber industry is, however, of a recent development of not more than a hundred years old. Rubber has been known to Amazonian Indians of the tropical South America since prehistoric time as they have used it for making balls, torches, and water tight container. The Mexican Aztecs played ball games with balls made of rubber. Europeans noticed this in the early part of the sixteenth century. Spanish settlers in 1600 used latex for smearing shoes, clothes, and hats, etc. for making them waterproof. *La Condamine* (1734) reported '*heve*' or '*jeve*' to be the native name for rubber in the Andean regin.

Later the rubber tree was given the generic name of *Hevea*. The name 'Rubber' was given to 'caoutchouc' due to it's ability to rub off pencil marks on paper, discovered by Joseph Priestley in 1770 in England. *Macintosh* in 1823 discoverd that rubber was soluble in naptha. This initiated the waterproofing of cloth in places where fresh latex was not available. Complete revolution of rubber industry occured when Goodyear invested vulcanization sixteen years later in 1839. The earlier trouble with that it became soft with heat and brittle with cold. During vulcanization rubber combined chemically with sulphur tear. This most significant discovery and the development of pneumatic tier has established rubber as one of the foremost items of commercial importance.

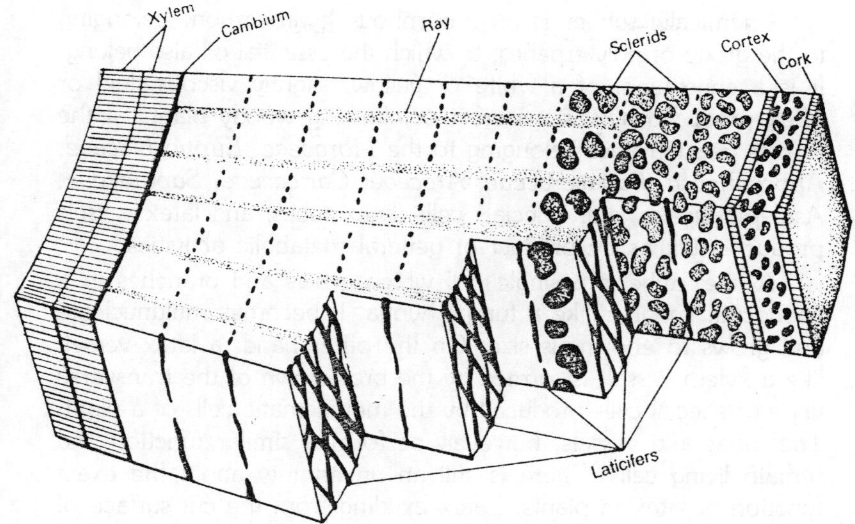

Fig. 10.1. A cross section of a hevea stem shows the location of the laticifers.

In Brazil willed trees of *Hevea brasiliensis* growing in the Amazonian rain forest were cut down to collect rubber. When its great value was realized by the Brazilian government it banned the export of seeds. However, some seeds were smuggled out to the Royal Botanic Garden at Kew, in London by an Englishman 'Farris' in 1873.

The seedlings were later shipped to Royal Botanic Garden at Calcutta but failed to survive there and were lost. Two years later, in 1875, Sir *Henry Wickham*, under orders from the British government collected about 70,000 seeds along the banks of the Amazon and shipped them in baskets labelled 'delicate specimens for Queen Victoria's garden at Kew'. 1,900 seedlings were later shipped to Sri Lanka which laid the foundation of the vast rubber plantations of Sri Lanka, Malaya and Indonesia. Liberia west Africa also have well established number plantation on a large scale. It has not, however, been possible to establish plantations in Brazil, the home of rubber, because of certain indigenous diseases and other factors. Malaysia, Thailand, Indonesia, Sri Lanka and India together produce 90 per cent of the world's natural rubber. In India–Kerla, Tamil Nadu and Karnataka are rubber-growing states.

Formation of Latex in Plants

Chemically rubber is an amorphous hydrocarbon, belonging to the group of polyterpenes, to which the essential oil also belong. It is a constituent of a white or yellow, slightly viscous fluid or *latex* produced by various erect or climbing woody plants of the torpics or subtropics belonging to the *Moraceae*, *Euphorbiaceae*, *Apocyanacea*, *Papaveraceae*, *Araceae*, *Caricaceae*, *Sapotaceae*, *Asclepiadaceae*, etc. Specials cells (latex tubes) and latex vessels produce latex as by product of general metabolic activities.

A latex tube is a single cell which grows and branches in a coenocytic manner like a fungal hypha. It becomes multinucleate and grows to enormous size. On the other hand, a latex vessel, like a xylem vessel, is formed by the breakdown of the transverse walls of special cells produced by the meristematic cells of a plant. The tubes and vessels, however, perform a similar function and remain living cells. There is still an uncertainty about the exact function of latex in plants. Latex exuding from the cut surface of a stem, leaf or root seal off a wound. It may also act as a fluid reservoir, in transport of material and for protection and nutrition. Different species of plant contain different composition of latex and only the latex of some dicotyledons yield rubber.

The rubber occurs as minute microscopic particles floating in the watery latex. There is a protective layer of nitrogenous substance around the particles to prevent their coagulation. Besides the particles of rubber, latex may contain varioius acids, mineral salts, sugar, starch, proteins, enzymes, alkaloids, resins, gums and oil etc. An account of the most important source of rubber, the Hevea or Para-rubber tree, *Hevea brasiliensis* is given below.

Hevea Rubber

Hevea brasiliensis belongs to the family *Euphorbiaceae*. It is native to tropical South America (Amazon Valley) where it grows wild in the hot damp forest. It also grows wild in eatern Peru, Bolivia, Colombia, Brazil, etc. It has been introduced in Liberia, Sri Lanka, Malaysia and Indonesia. 97 per cent of the total world production of rubber is constituted by the last three countries mentioned. The plant grows well within a temperature range from 75 to 90°F and a rainfall of 80–120 inches. The para-rubber tree grow to a height of 60 to 150 ft. The tree trunk grows to a thickness of six feet. The tree bears compound leaves of three

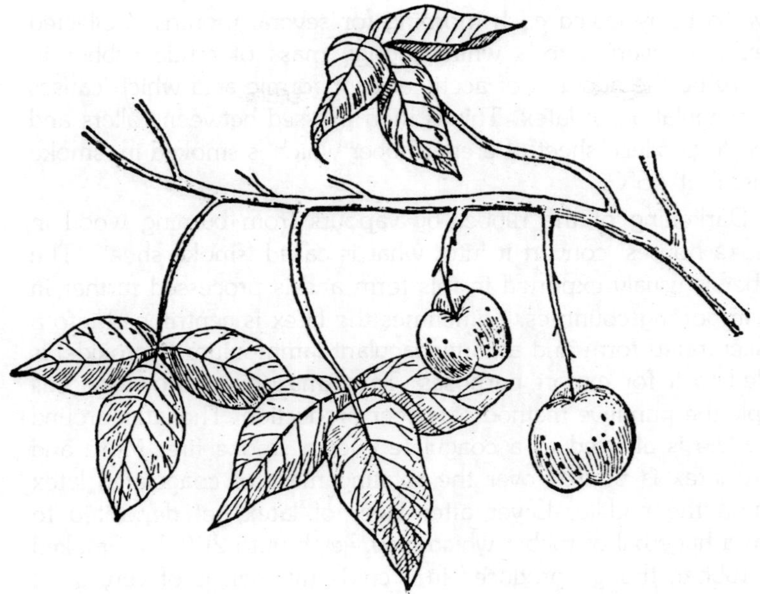

Fig. 10.2. Leaves and fruits of para-rubber tree.

leaflets. The flowers are small an dare borne in inflorescences. The fruits are three lobed with each lobe bearing one seed. About 50 per cent of a dark red drying oil is contained in seed.

Tapping of Rubber

Latex of para-rubber tree contains about 30 per cent dry rubber. The Latex is obtained from the numerous latex vessels of the bark which are arranged in concentric rings alternating with rings of phloem. The inner bark contains more vessels than the outer bark. The bark is cut in such a manner that the delicate growth layer of cambium is not damaged. To obtain maximum yield tapping cut is made from upper left to the lower right at an inclination of 30° since latex vessels run spirally to right at an angle of 30°.

Tapping begun in six-year old trees. First a vertical cut is made in the bark in the lower portion of the tree. Several short downward spirally running incisions which extend either one-third or one-half of the circumference of the tree are made in the bark with a special knife. Sometime a 'V' cut is made. The latex which flows for several hours in the forenoon is collected in a cup. The flow of latex stops by midday and. therfore, cut surfaces have to be renewed each morning for several months. Collected latex is converted to a white spongy mass of crude rubber in factory by the addition of acetic acid or formic acid which causes the coagulation of latex. This is then pressed between rollers and dried to produce sheet of crepe rubber which is smoked in 'smoke houses' at 45°C.

Darkening of the rubber by vapours from burning wood in 'smoke houses' convert it into, what is called 'smoke sheet'. The rubber is usualy exported in this form and is processed further in the importing countries. Sometimes the latex is centrifuged into a concentrated form and an anticoagulant (ammonium hydroxide) is added to it for export purposes. The Amazonians, however, still emply the primitive method of rubber production. The latex around a paddle is allowed to a coagulate over a smoke fire. More and more latex is poured over the rotating mass of coagulated latex around the paddle. Layer after layer of latex get deposited to form a huge bal of rubber which may weigh upto 200 lbs. Smoked ball rubber, though produced in a crude manner, is of very good quality.

In the recent past considerable attention is being paid for the improvement of rubber plantations. A high yielding plant is obtained by inserting a bud from a high yelding clone on a seedling having vigorous growth. Similarly the plant is then made disease-resistant by inserting another bud to the top from a disease-resistant clone. Thus, a tree with a vigorous rootstock, high yielding trunk and disease-resistant top is obtained as a result of double budding technique. In the recent past chemists have been able to produce synthetic rubber. Catalytic polymerization of butadiene and styrene in soapy water gives a latex which can be coagulated with acid and salt, filtered, and dried to yield synthetic rubber. The synthetic rubber is now greatly improved. Mixture of natural and synthetic rubber, now-a-day is used to manufacture most of the automobile tyes.

Other Sources of Rubber

Palay Rubber

The rubber is obtained from *Cryptostegia grandiflora* and *C. madagascariensis* of the *Apocyanaceae*. They are native to Madagascar on the East African Coast. They are also grown in Haiti.

Assam Rubber

Assam rubber or India rubber is obtained from *Ficus elastica* of the *Moraceae*. It is native to northern India and Malaysia. The rubber is of poor quality.

Panama or Castilla Rubber

It is obtained from *Castilla elastica* which is native to Mexico and Central America. Upto 1850 is/was of greater importance than the hevea rubber and was the chief source of rubber. The plant belongs to the *Moraceae* family.

Dandelion Rubber

It is obtained from the Russian dandelion (*Taraxacum koksaghza*). Tuberous roots of the plant is the source of latex.

Guayule Rubber

It is obtained from guayule (*Parthenium argentatum*) of the *Compositae*. It is native to the United States.

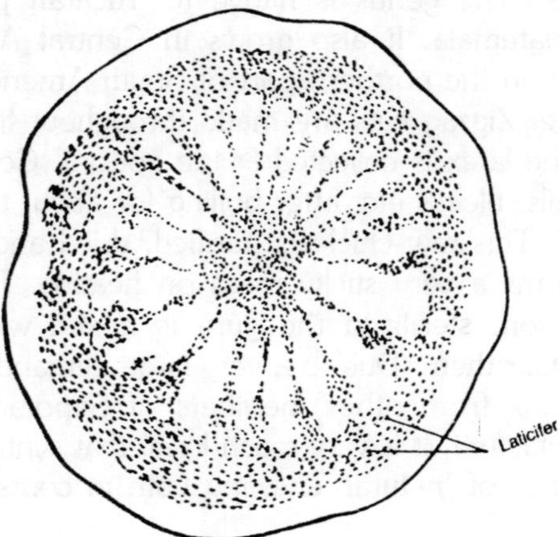

Fig. 10.3. A cross section of a guayule stem showing the distribution of laticifers.

Ceara or Manicoba Rubber

It is obtained from several species of *Manihot* which belongs to the *Euphoribiaceae*. The important rubber-yielding species are M. glaziovii, the Ceara manicoba of northeastern Brazil, *M helptaphylla*. *M. dichotoma*, etc. The plants are grown in Sri Lanka. India and many other tropical countries.

Landolphia Rubber

The plant are woody vines of Africa. They belong to

Landolphia of the *Apocyanacea*. The important species are *L. kirkii*, *L. heudeloti* and *L. owariensis*.

Use of Rubber

1. Rubber is extensively used in the manufacture of boots, shoes, hose tubes, mats, belting, raincoats and other waterproof materials, toys, insulated wire, etc.
2. Crude rubber is vulcanized with 30 per cent sulphur to produce hard rubber, which is used in telephones, radio parts, surgical appliances, etc.
3. Manufacture of tyres and tubes of automobiles, aeroplanes, bicycles, etc. consumes three fourth of the crude rubber.

Other Latex Products

Chicle

Chicle is a latex product like a balata. Chicle is the name of the gum which is obtained from the latex of *Achras zapota* of the *Sapotaceae*. The genus is native to Yucatan peninsula of Mexico and Guatemala. It also grows in Central America, the West Indies, and in the northern part of South America. If is also a wild large tree. Zigzag cuts are made throughout the length of the trunk and the latex is collected at the bottom. Collected latex is made into balls, blocks etc. after boiling for about two hours in large couldrons. This raw chicle is purified, dried and powdered. The powder forms a very sticky mass on heating. Finally, after cleaning filteration, sterilized the gum is mixed with different materials to flavour them. Due to a very large demand of chewing gum inferior latex from other members of Sapotaceae is also used in its manufacture. It is very much likely that synthetic plastics will take the place of 'natural' chewing gum in course of time.

Gutta-Percha

Gutta-percha is a non-elastic rubber formed from a greyish white latex which obtained from some members of the family Sapotaceae. The latex is chiefly obtained from *Palaquium gutta*. The plant is native to Malaya and grows in Malayasia, Indonesia, Philippines, and other tropical countries. The latex is obtained either by cutting the trees or by removing barks and making cuts. The latex is collected on banana leaves or is coconut shells. It coagulates immediately. Gutta-percha is resistant to salt water and pliable, thus it is widely used for construction of submarine cables. It is a very poor conductor of electricity and, therefore, it is used

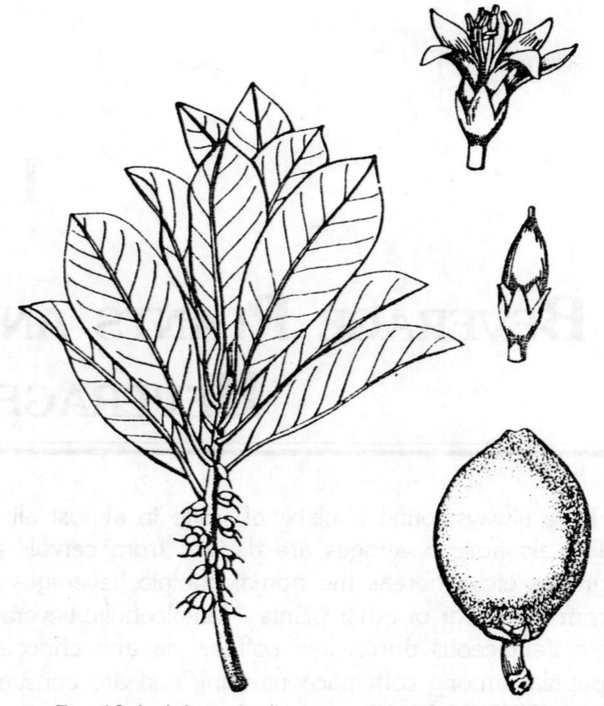

Fig. 10.4. A branch, flower, and fruit of Palaquium gutta.

for insulation purposes. It is also used in the manufacture of pipes, golf balls, telephone receivers, adhesives, toys, supports, etc. It protects wounds and is used in dentistry.

Jelutong

Jelutong is a gum obtained from the latex of a Malayan tree *Dyera costulata*. It is very much used as a substitute of chicle. For sometimes it was also used as a source of rubber. In dry regions gums are familiar secretions. They are commonly used as adhesives. Besides, they are used in printing, paints, candies, medicines, and for finishing textiles and sizing of papers.

Balata

Balata like gutta-percha is a non-elastic rubber. It is obtained from the latex of *Manilkara bidentata* earlier called (*Mimusops balata*). This large tree is indigenous to South America and Trinidad, where it grows in a wild state. The uses of balata are similar to those of gutta-percha. It is specially useful in the manufacture of machine beltings.

11

Beverage Plants and Beverages

Beverages have always found a place of pride in almost all the societies. The alcoholic beverages are derived from cereals and fruits like grapes etc. whereas the non-alcoholoic beverages are prepared from a number of other plants. Non-alcoholic beverages consists of caffeinaceous drinks like coffee, tea and chocolate, these are popular among caffeinaceous drink and are consumed all over the world. The three non-alcoholic beverages mentioned above are described below:

Coffee (*Coffea* spp.)

Origin and History

There are 90 species of coffee of which four are important from the point of view of beverages. Most of the species are native of Africa and Madagascar. *Coffee arabica* is the only polyploid (tetraploid, 4x = 44) species. The *arabica* type is indigenous to south-western Ethiopia in the north-eastern Africa. The *arabica* type spread to Yemen in the fourteenth or fifteenth century. From Yemen it first spread to the Malabar coast of India and then to Sri Lanka and Java (Indonesia). It was introduced in Arabia only 500 years ago. It reached to piligrims and traders via Turkey to Europe in the sixteenth and seventeenth centures. The plant was introduced in Sri Lanka, India and Java by 1700, in the West Indies in 1720 and in Brazil in 1770. It was carried to U.S.A. in the nineteenth century.

Canefora (robusta) coffee, a diploid species, has its centre of origin in central and western equatorial Africa and in Madagascar. It was introduced (20 plants) into southeast Asia (Indonesia) about 80 years ago from Zaire.

Canefora coffee was found there to be a very good substitute for the arabica type which was earlier devastated completely by the coffee leaf rust disease (causal organism = *Hemileia vastatrix*). USA is the greatest coffee-consuming country. The world production of coffee is more than 5 million tons, out of which South AMerica, particularly Brazil produes 4 million tons of *arabica* type. Arabica types constitues 80 percent of the total production of coffee.

Botanical Description

Coffee plant belogns to the family *Rubiaceae*. The three different kinds of coffee are follows:

(i) *C. robusta* or *C. canephora* (Congo coffee).

(ii) *C. liberica* (Liberian coffee).

(iii) *C. arabica* (Arabian coffee). It is the source of best quality coffee and accounts for 90% of the total world production of coffee.

The coffee plant is a shrub or small tree with a height of 15 to 50 ft. The leaves are oppositely arranged having dark green. Flowers are white axillary and appear two or three times a year. The fruits are reddish two-seeded or occasionally one-seeded drupes. The mesocarp of the coffee berries while endocarp is thin membranous. The leaves and the fuits contain caffein (1 to 2%). The green seed contains 34% cellulose, 8% chlorogenic acid, 10 to 13% oil, 7% sugar, 14% protein, 12% water and some other components.

Cultivation

The grown of the plant is better on highlands of a height of 4,500 ft. or so. On the other hand they can grow on plains as well as on high altitudes upto 6,000 ft. They like moderate temperature and frequent mists of the hills (moist climate). They grow best on fertile, well drained soils. The seeds are sown first in seed beds. The seedlings are transplanted in rows about 9 ft apart. The plant bears fruits from third year and continues to do for about 25-30 years. Planting, pruning and picking of the berries are all done by hand. Picking is done when the fruits are fully ripe. They may be stripped off or be allowed to fall down on the ground. They are then winnowed by throwing them in air.

Fig. 11.1. A—A branch of coffee with clusters of berries in the leaf axils; B—a cluster of jasminelike flowers; C—an individual flower in D—cross section; E—the position of the two seeds in the fleshy fruit; F—the ventral (flattened) side of the seed; G—a cross section of the whole berry showing the component parts.

Preparation

Coffee can be prepared by two method: (i) Dry method, (ii) Wet method. In the dry method the coffee berries are dried in the open. The dried skin and pulp are separated from the seeds by mechanical means, or by pounding them in a mortar. In the wet method the fruits are taken to the flotation tanks where the ripe berries float and the debris including the green berries sink.

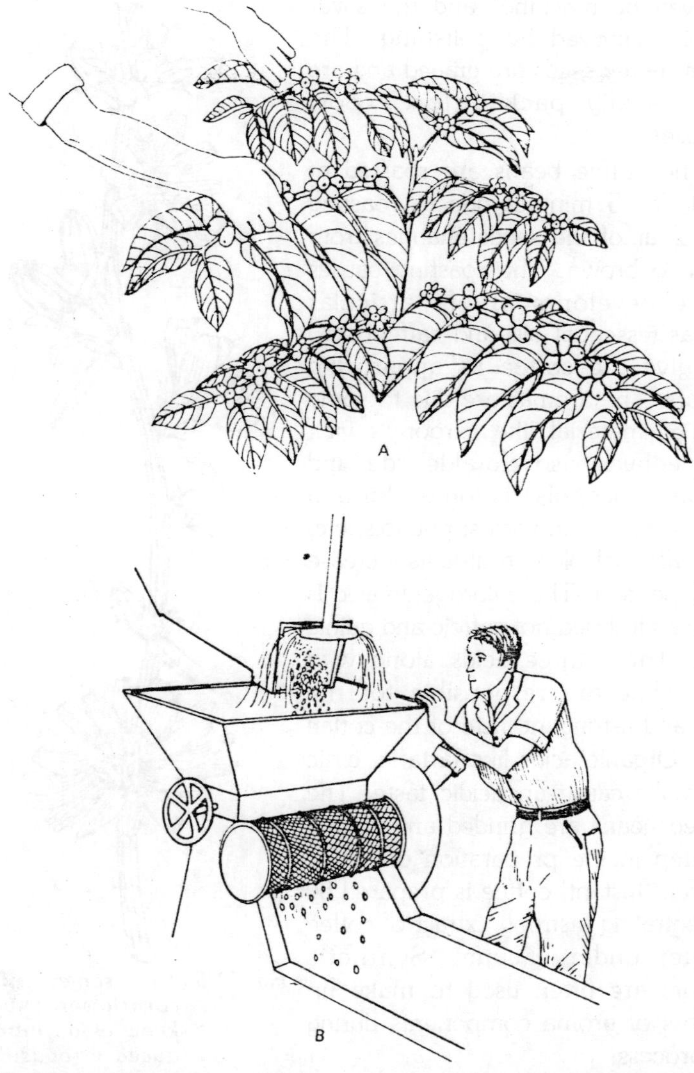

Fig. 11.2. Steps in the processing of coffee. A—Selective hand picking of only the red, ripe berries produces coffee with the finest flavour; B—Washing the berries.

The seeds are removed from the pulp by means of pulping machines. Parts of the pulp sticking to the seeds are removed by controlled fermentation over a 12 to 24-hour period. Seeds are then thoroughly washed in water and are sprayed on mats in the open for drying. The brittle, dry husklike parchment (endocarp) is removed by machines and the silver skin is removed by polishing. The greenish-grey seeds are graded and are then usually packed for export purposes.

The coffee beans are roasted at 500°F for 5 minutes due to roasting the colour of the seeds changes from green to brown. The roasting causes in the development of particular aromas (essential oils and caffeol etc.) that give the coffee its appertizing flavours. The aromas are due to some volatiles material like carbonyls viz., acetaledhyde, isovaleraldehyde and acetone, alcohols, ketones, furfuryl compounds, amino acids, phenols, etc. The water soluble constituents increase to 35 percent. The chlorogenic acid is partly hydrolysed not caffeic and quinic acids. These three acids along with trigonelline are responsible for the bitter and astringent taste of the coffee brew. Organic acids like tartaric, cirtic and malic cause an acidic taste. The roasted beans are grinded and is the last step in the preparation of coffee powder. 'Instant' coffee is prepared by concentrating a strong extract of coffee powder under vacuum. Synthetic flavours are often used to make up the loss of aroma components during the process.

Fig. 11.3. The same wild cornflower we know as an introduced roadside weed provides the roots that are roasted and ground to make chicory.

Roasting and grinding are done only few days before consumption of coffee otherwise it loses the flavour and the oils become rancid. The coffee from different sources are tasted by a 'coffee taster' for blending them into a particular kind. The roasted coffee beans havae 0.5 to 1.5% caffein, an alkaloid.

Caffeine

About .8% to 1.8% caffeine present in commercial coffee. Instant cofee contains 4.4% caffeine. One cup of coffee roughly contains 100 milligrams of caffeine. The solubility of caffeine in water is only 2.2%. the caffeine after assimilation enters the blood streams and reaches all parts of the body within five minutes. It is, therefore, a potent drug, which is both beneficial and harmful depending upon the dose and frequency of intake. The amount of caffeine which gets distributed in various tissues is approximately in proportion to their water contents. Caffeines quickly decomposed into 1,7-dimethylxanthine in the tissue

The demethylated products appear in the blood. It, however takes about one week for the complete clearance of caffeine through urine. Caffeine acts on the central nervous system. Sensory stimuli and motor activity act at a quick rate. It prevents drowsiness and fatigue. It relaxes the muscles of the bronchial tubes. It stimulates medullary, respiratory excitation of lower motor centers. It has an over-all effect on the cardiovascular system. It helps in the constriction of the muscles of the heart and dilates the coronary, pulmonary and general systemic blood vessels resulting in increase in the rate of flow of blood to evey part of the body including brain.

Hazards of Caffeine

There are more than 10,000 scientific reports about the undesirable effects of caffeine. The alkaloids causes several health hazards. The addicition to coffee is such that stoppage of the drink results in headache, irritability and sometimes nausea. Luckily addiction to caffeine can be got rid off in a few days. Excess of caffeine can spoil nervous of a person. The "restless legs syndrome" of psychiatric patients has now been partly attributed to the excess intake of caffeine. Excess of caffeine causes over-excitement, insomnia, mild delirium, sensory disturbanes such as ringing in the ears and flashes of light, restlessness, fluttering of heart, several heart diseases, peptic ulcer, turmours and birth defects takes place due to excess intake of caffeine. Caffeine is more harmful to children since they cannot eliminate it as fast as an adult can.

The caffeine containing cola drinks are a great hazard for the children. The cola containing cold drinks contain 35 to 55 milligrams per (360 ml) bottle. Administration of very high concentration of caffeine during pregnancy has been shown to cause foetal deaths and malformation in mice. It is therefore, advisable to avoid these hot and cold drinks during pregnancy particularly in the early stages. The soft drink industry should remove caffeine from drinks for the safety of children. It should be noted that the morning cup of coffee or tea gives the feeling of well being with greater concentration and intellectual activity followed by, sadly enough, the effeciency of coffee reduced after morning.

Uses of Coffee

1. It has therapeutic value in the treatment of bronchial asthma as it relaxes smooth muscles of bronchi.
2. It is diuretic i.e., causes increased production of urine.
3. The waste produces, pulp and parchment are used as fertilizer, and fuel. They are also used for the manufacture of a plastic material, cafelite, used for insulating purposes.
4. In Sumatra coffee leaves are used like tea leaves to prepare beverage.
5. Alkaloids are stimulative which makes the coffee a much wanted beverage. It relieves fatigue and stimulates the physical and mental activities.
6. It is used in the treatment of cases of poisoning caused by alcohol and morphine.
7. Caffeine is used to several drugs. It relieves headache by its vaso-constrictor effect on blood vessels.
8. It is given along with ergot to relieve pain of migraine.

TEA (*CAMELLIA SINENSIS*)

Origin and History

Tea is the most popular beverage. It is believed to be native to southwestern China, northeastern India (Assam) and adjoining areas of upper Burma. It is found mostly in a wild state in these areas. Tea has been cultivated in China since early times and is believed to have spread from its primary centre of origin i.e. lower Tibetan mountains or central Asia to a secondary centre the sources of Irrawady. From the latter site tea spread in three directions giving rise to three types of tea viz., China, Assam and Cambodia. It is assumed that the small-leaved to be a secondary centre of origin. *Wight* (1962) has named the two important types of tea plants as *C. sinensis* (Chinese) and *C. assamica* (Assamese).

The Cambodian type is classified as a subspecies of *C. assamica*. Tea was carried to Japan in 1000 A.D. The Dutch introduced it into Europe in 1610 A.D. It reached London in 1664 and Boston (U.S.A.) in 1714. The cultivation of tea is world wide in the tropical and temperate regions extending from 40°N in Russian Trancaucasia to 33°S in Argentian.

Tea Growing Regions

The major tea-growing regions are all situated in Asia, viz., China, Japan, Formosa (Taiwan), Java, Sumatra, Bangla Desh, India and Sri Lanka. Commercial tea plantations were started in India. Sri Lanka and Indonesia in the nineteenth century. India and Sri Lanka are the world's largest two tea producers and together account for two-thirds of the world's total exports of tea, about 210 million kilogrammes each (Rs.200 crores in foreign exchange). The principal tea growing states are Assam, northern districts of Bengal, Kerala and the NIlgiri hills. Some other tea-growing areas are Ranchi, Dehradun, Kangra and Kumaon districts.

Botanical Description

Tea belongs to the family *Thaeceae*. The tea plant is an evergreen tree or shrub which can grow to a height of 30 to 50 ft. It is however, not allowed to grow to that height. Under cultivation it is prunned down to a small height of 2 to 5 ft so that it looks more like a shrub. It bears alternately arranged leaves which are elliptical in shape and serrated in margin. They are glabrous. The young leaves are however, pubescent. Large number of oil glands are present in the leaves. Old leaves are bright green coloured. They may be up to 12 inches in length. White and fragrant axillary flowers may occur singly or in clusters of few. The fruits is a 1 to 4-celled capsule, each chamber containing 1 to 3 seeds. Tea has about 1,000 varieties.

Ecological Factors

The tea plant grows well in tropical and warm temperate climates. It requires plenty of rainfall. A minimum annual rainfall of 60 inches is essential for its proper growth and development. The rainfall should be evenly distributed throughout the year because uneven distribution of rainfall result poor growth. Prolonged drought is extremly harmful and damages the tea crop. Whereas the Assam and Bengal plantations get plenty of rainfall 50"-150" and 100"-200" respectively the tea farms of Ranchi and Dehradun are poor in growth due to uneven distribution of rainfall. High temperature accompanied with rainfall is good for tea plant.

The optimum monthly temperature between 70°-90°F is excellent. Temperature below 65°F is not good for its growth. Frost damages the crop. Shady conditions are better than exposed situations. Though tea has been cultivated on hilly tracts up to an altitude of 7,000 ft. It is usually grown in areas lying between 3,000 and 4,000 ft. above the sea level.

It can, however, be successfully grown on flat plains provided the soil is well drained. Some of the finest tea plantations in Assam are on levelled grounds, lying between an altitude of 50-400 ft. The tea plant requires well manured and humus rich deep soil but it can grow on poor soils as well. The soil usually ranges from light sandy to stiff clayey types. Acidic soil (pH 5.2 to 5.6) is good for proper growth to the tea plant. The soil must be well drained.

Cultivation

Nursery

Tea is usually grown by sowing seeds. Carefully selected seeds are sown in well prepared nursery beds. The nursery beds are usually shaded. Seeds are sown 4"-8" apart. The nursery is manured till the seedlings are 6-9 months old.

Preparation of land

The land where the tea seedlings are to be planted is also prepared. Land previously occupied by forest is preffered for tea plantation. It is first completely cleared of trees and is then ploughed to remove the roots etc. The steep slopes of the hills are contour terraced. Drains are usually dug along the contour of the slope to prevent erosion of soils and waterlogging of the low-lying areas. The size and number of drains vary with the stepness of the land, the nature of the soil and the amount of rainfall.

Planting of seeds

The seedlings are planted in holes dug 3 ft. to 4½ ft. apart depending upon whether land is terraced or is flat. The holes are filled by the mixture of organic manure and soil. The plantations are shaded by growing leguminous plants spaced about 50 ft apart.

Weeding

A lots of methods used to keep the tea plantations free from weeds. The weeds are prevented from growing by scraping the soil surface and also by deep cultivation.

Pruning

Pruning is done for three purposes: (i) To include vigorous growht of the tea plant for development of new shoots; (ii) To ensure a continuous supply of tea leaves to the tea industry, and (iii) To keep the height of tea plant reasonably low so that the leaves can be plucked by hand. There are two types of pruning—(i) light pruning and (ii) heavy or medium prning. When the plants are 2-5 years old, during their slow growing period, the first few prunings are done. The light pruning may be done annually or biennially. Heavy pruning is resorted to after ten years or so when the plant is cut back to the ground level so that the suckers replace the old bush.

Plucking

Plucking is done when the plant is 4 to 5 years old. The quality of tea depends upon the age of the leaves. The tannin content of the leaves is variable with their age. The young leaves have more tannin make better tea than old leaves with less of tannin. The percentage of tannin in the tea 'flushes' is as follows: bud—25, 1st leaf–28, 2nd leaf–21, 3rd leaf–14, stalk between 2nd leaf and bud –12, and stalk between 2nd and 4th leaf–6. From above data it is clear that the terminal bud or 1st leaf gives excellent quality of tea. The bud is usually plucked along with 1st and 2nd leaves. Inclusion of more than two leaves gives 'coarse' tea. Sometimes al the 4 or 5 leaves of a 'flush' are collected separately to prepare 4 or 5 different qualities of tea. Plucking is done by hand by women and children.

Plucking may be done at intervals of 7-10 days since the new shoot develops very quicky. In cold climates there may not be any plucking during winter months because the growth of the plants almost comes to a stop. In hotter regions 25 or 30 pluckings can be made in a year. The bushes of tea may surviving for as many as 200 years but the yield decrease substantially after the plant is 40 to 50 years old.

Preparation of Tea

Commercially there are four kinds of tea are prepared: *black* tea, *green* tea, *oolong* tea and *let-pet* or *leppet* tea.

Black tea

Black tea is by far the most important of all. It includes four steps in the preparation of black tea.

Withering. Harvested fresh leaves are sent to the factories where they are withered for about 24 hours on well ventilated indoor shelves or to pass over them in the houses which have open sides. The withering results in lowering of the water content of the leaves which become soft and flaccid. Withering is usually omitted in the case of green tea preparation.

Rolling. Withered leaves are sent to the rolling machines, which in several rolling and breaking operations damage the individual leaf cells so that the juice and enzyme contents are released and the whole rolled mass gets smeared with them. This facilitates fermentation. The polyphenols called *catechins* (25%) are oxidised into *theaflavin* and *thearubigins* by the enzyme polyphenol oxidase. They give the characteristic taste, briskness, colour and strength to the brewed tea.

Fermentation. Very high temperature causes over fermentation which spoils the quality of tea. So high temperature avoided and the leaves are allowed to ferment for 2-6 hours. The colour of leaves changes from green to bright reddish coppery colour and develop pleasing aroma during fermentation. While green tea did not require fermentation.

Drying or *Firing*. The fermented leaves are conveyed through air chambers at 130°F or so for 30 to 40 minutes and at 192-200°F for a while in the end. The leaves coming out are left with only 3 to 4% moisture.

Grading. The prepared tea is graded into various kinds. The leaf grades (orange, pekoe) give higher tea than 'broken'grandes (broken orange pekoe, broken pekoe etc.) The broken grades are used for quick preparation of tea particularly in restaurants, canteens etc. Broken grades can be packed in larger amounts in a container as compared to leaf grades.

Packing. Packing is done in plywood boxes (24" × 19" × 19") which are lined with aluminium foil or parchment paper to protect them from moisture. The small tea packets are lined with metal foil.

Blending. Blending is an art by which the different qualities of a particular tea brand produced in a tea estate are blended together to give a uniform product.

Green tea

The fermentation and withering is not required in the preparation of green tea. Its preparation includes heating, rolling, and drying. The latter two processes are similar to those of black tea preparation. Heating may be done either by steaming or by pan firing. The heating is completed within a few minutes so that the green leaves are without any sign of scorching or reddening.

Oolong tea

An exclusive produce of Taiwan (Formosa), oolong tea is favourite of Americans. It is manufactured in the same manner as that of black tea except that the oolong tea is allowed only a light fermentation. The semi-fermented tea has a characteristic flavour more because of the particular variety of tea grown under a particular set of soil and climatic manufacturing process. A similar semi-fermented tea is the scented jasmin tea. It is prepared by allowing the fragrant jasmin flowers to dry along with the tea leaves. The flowers are later on removed.

Brick tea

Usually its manufacturing is done in China from where it is exported to Tibet and U.S.S.R. Brick tea is prepared from coarser leaves, twings and remains left over from the preparation of black tea viz., dust, stalks, etc. They are steamed over a boiler. Firmness is provided by mixing a little rice paste before they are pressed into moulds.

Let-Pet or Leppet tea

In Burma boiled or steamed green leaves of tea are preserved in pits. They are used more as pickles or vegetables than as beverages.

Uses of Tea

Caffein like alkaloid called theine (2 to 5%), a volatile oil, and 14 to 28% tannin present in the tea. While tannin does not add to the drinking quality, the volatile oil and the alkaloid theine which dissolve quickly in hot water give a refreshing and stimulating beverage. It relieves the body fatigue. Excessive drinking of tea is considered to be harmful for the digestive system.

COCOA AND CHOCOLATE (CACAO=*THEOBROMA CACAO*)

Origin and History

Cocoa and chocolate are the most nutritious of all beverages. The sources of Cocoa and Chocolate is the seeds of the cacao tree, a native of South America. It was a favourite beverage of the Aztecs, who made a cold, frothy liquid seasoned with peppers

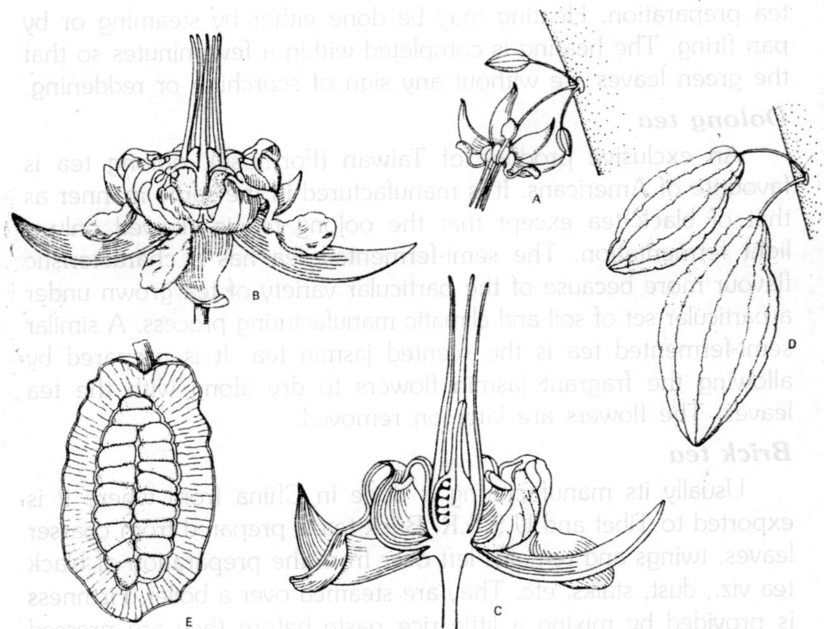

Fig. 11.4. Flowers and fruits of a chocolate plant. A—A cauliflorous flower; B—Flower greatly magnified; C—Cross section of a flower; D—Fruits borne directly on the trunk; E—Cross section of a fruit.

and spices and called it 'chocolate'. Spaniards introduced it into Europe in the sixteenth country, when along with coffee it became a very popular beverage.

Initially it was grown in Brazil, Equador, and in other neighbouring regions but today it is also grown extensively in Africa (Ghanna, Nigeria, etc.) Java and Sri Lanka. Two-thirds of the world producition of cacao is from Africa (Ghana and Nigeria) while the rest is obtained from South and Central America and the West Indies. Indonesia and Sri Lanka produce a very small fractio of the crop. The crop after processing is mostly exported to the temperature countries which are the leading consumers. Europe consumes more than 50 per cent and America about 40 per cent of the world's production cacao.

Botanical Description

The cocoa tree belong to the family *Steruliaceae*. It is a branched tree with small height (15 to 25 ft) and bears long leaves which may be up to 12" in length. Flowers appears on the trunk of longer branches of the tree while fruits on short stalk. Fruits are pod like and each pod bears 49 to 60 large seeds known as beans. The fruits start developing in 4 to 5 years old trees. There is no particular flowering season and the plant produces flowers throughout the year. Several crops of fruits are, therefore, harvested in a year.

Ecological Factors and Cultivation

A hot and moist climate of the tropics are required for proper growth and therefore, it grows well within 20° if the equator. It, however, needs protection from direct rays of sun and also from strong winds. This is why cocoa is grown along with catch crops and permanent shade trees. Well drained, moist, deep alluvial soil is necessary for its cultivation. It is sensitive to low temperature and high attitudes. It cannot grow below 60°F and altitudes beyond 2,500 ft. The seeds are sown in prepared seed beds. The seedlings are transplanted 4 ft apart.

Preparation

The pods are cut with knife and the pulp of the fruit and the seeds are separated from the pods.. They are dried in the sun or are allowed to ferment sufficiently. The seeds are washed to remove the disintegrating pulp. They are then dried and polished. The sorting and roasting is done at high temperature in iron drums.. Roasting results in increase in the fat (50%) and protein contents, decrease in tannin content and in the development of characteristic aroma, as a result of complex chemical reactions occurring in the cotyledons. The small broken pieces of seeds are called *nibs*. The nibs are made into an oily paste called 'liquor'. Removal of oily cocoa butter and making powder of the residue results in cocoa. If the cocoa butter is left in and sugar is added chocolate is formed.

Uses of Cocoa and Chocolate

1. The 'cake' gives caffein like alkaloid theobromine which is used in soft drinks "colas" and for other purposes.
2. The cocoa shells are used as a cattle feed and as fertilizer. They can also be used as a beverage and as an adulterant of cocoa and chocolate.
3. They are an excellent beverage. The high fat content of the bean is reduced by half (to 25%). the rest of the extracted fat is sold as cocoa butter.

4. It is used in the finest of chocolates, candies and cakes etc.
5. Cocoa butter is a very fine edible fat. It is used for a number of preparations.
6. They contain considerable nurtrient material in the form of fats, proteins and carbohydrates and, therefore, they are nutritious.

Other Caffein Containing Beverages

Mate. It is also Paraguay tea. It is obtained from the leaves of various species of holly, chiefly *Ilex parayuariensis*.

Gaurana. It is prepared from the seeds of *Paulinia cupana* of the Amazon valley.

Khat. It is obtained from the leaves of *Catha edulis* of north-eastern Africa.

Cola. The seeds of *Cola nitida* (*C. accuminata*) is the sources of cola. The seeds are usually called cola or kola nuts.

Cassine. It is a tea like beverage. It is obtained from *Ilex vomitoria* of Mexico.

Yoco. It is obtained from the bark of yoco (*Paullinia yoco*) of Peru, Equador and southern Colombia.

Other Beverages

Other non-alcoholic beverages are normally comes under the group of *soft drinks*. They are usually *juices of fuits* like orange, lemon, apple, pineapple strawberry raspberry, etc. The fruit juice is pasteurized and is mixed with preservative before bottling or canning it. The fruit juices are rich in sugars and are, therefore, good sources of energy. Synthetic flavours are widely used in commercial products. Other soft drinks are *malt beverages*, *ginger ale*, *sarsaparilla*, *cola beverages* and *soda water*.

Alcoholic Beverages

Alcoholic beverages are common drinks of the rich and the poor from times immemorial. Alcohol has deleterious effects on the human systems and excessive drinking is extremely dangerous. Alcoholic beverages can be studied under the two: (i) *fermented beverages* and (ii) *distilled beverages*. The fermented beverages include *wine*, *beer*, *hard cider*, *root bear*, *mead*, *sake*, *palm*, *wine*, *pulque* and *chicha*. The juice of grapes (*Vitis vinifera*) is fermented to obtain wine.

Wine is prepared chiefly in the European countries of France, Germany, Italy, Hungary, Spain and Portugal. Beer is prepared from barley, rice, maize, etc., by first converting the starch into sugars (*malting*) and then fermenting them (*brewing*). Beer consists of 3 to 8 per cent of alcohol, sugars, dextrin, phosphates and various proteins. Hard cider is prepared from apple juice. Juices of palm, gaguey, banana, sugarcane, yucca sorghum, cassava, sweet potato, pineapple, etc. are also used for preparign beverages. The distilled beverages include *whisky*, *brandy*, *rum*, *gin*, etc. The fermentation of unmalted or malted cereals or potatoes give a mash which is distilled several times to yield 'high wines'. The high wine is mixed with water to yield whisky. The world famous Scotch whisy is prepared from barley. The famous Russian whisky Vodka, is prepared from fermented wheat mash. Brandy is distilled from wine. It has as high alcohol content of 65 to 70 per cent. Rum is distilled from sugarcane and molasses. It contains about 40 per cen talcohol. Gin is usually distilled from a fermented mash of malt of barley, rye, etc.

12

Drugs and Poisons from Plants

In the previous chapter, we discussed drugs that are used medicinally to releive pain or control disease. A large number of same drugs can effect an individual's sense of perception by producing feelings of transquality, invigoration, or otherworldiness. Sometimes people want to escape from, or "alter," reality and thus seek out these drugs, but in excessive quantities, most are toxic. This chapter focuses on plant substances used because of their psychoactive properties, and on the overlapping group of poisons obtained from plants. The psychoactive drugs can be classified on the basis of the chemical nature of the compounds involved, the effects they produce, or thesource from which they are obtained. None of these classification schemes is perfect.

The psychological effects of the compounds involved sometimes differ between chemically very similar substances, and a classification based on chemical structure is rather difficult for a nonchemist to remember. It is difficult to categorises the psychoactive drugs into stimultants or depressants, hallucinogens or narcotics because many drugs can act in a combination of ways. For example, many of the so-called depressants, like the narcotic analgesics (opiates) and the solanaceous alkaloids, can also act as hallucinogens. Stimulants such as nitotine and cocaine will likewise cause vision if taken in sufficient quantities. The word narcotic itself a dangerous and filled with problems because some people use it to refer strictly to those drugs obtained from the opium poppy (as here), while others employ it for any habit-forming drug or even for any drug used illicity. One way of grouping psychoactive drugs by their primary effects, but since this is a botany text, we will follow yet another system and discuss drugs taxonomically in terms of the species from which they are obtained. Caffeine and alcohol these are wildly used. Psychoactive drugs are discussed in next chapter. These are obtined from a variety of plant sources and are economically and culturally important.

THE CHEMISTRY AND PHARMACOLOGY OF PSYCHOACTIVE DRUGS

Almost all chemicals thathave psychoactive properties contain nitrogen, and most belong to one of the classes of alkaloids. The most notable chemical exception is the active compound of marijuana, Δ-*trans-tetrahydrocannabinol* (THC). Some other alcohols (including ethanol) and terpenes have psychoactive effects and have been designated as hallucinogenic agents. First of all the psychoactive drugs absorbed into the blood stream and transported via circulation to sites where they can exert their effects.

Like medicinal drugs, psychoactive drugs can be taken orally, injected into the body, or absorbed through membranes such as those lining the nose, rectum, vagina, or lungs. Once the active compounds enter the bloodstream, they are transported to all parts of the body. A drug may exert its influence in only one place in the the body, but it is still present (at least initially) throughout the bloodstream because blood and the products it carries continuously circulate around the body. As the blood passes through the liver on its cycle, some drugs are degraded in the places where they act, and the breakdown products are picked up by the blood and transported to the kidneys for excretion. In this process of circulation, ending in degradation and excretion, that accounts for the initial "rush" followed by an enventual "wearing off" of a drug's effect.

Most of the psychoactive drugs acts on cell of the central nervous system (the brain and spinal column) and affect very badly as the blood flow to the brain is 10 times that to other tissues. Therefore drugs reach the brain more quickly than tissues. However, blood capillaries around the brain are more tightly packed that those elsewhere and are covered by sheaths that make passage of compounds from them into the brain difficult.

Thus the actual amounts of drugs that enter the brain are less than the blood flow to it might suggest. Once they have reached the central nervous system. Psychoactive drugs disturbed the natural interactions between nervous, or sensory cells. These cells transmit information by chemical signals.

Neurons release chemical substances called *neurotransmitters* in response to stimulation. Once released, these transmitters flow across the space, or *synapse*, between a transmitting neurons and a receiving neuron. Specific sites on the receptor recognize the transmitted compound and bind briefly with it. Once the compound has been accepted, it triggers a response in the receptor neuron. Among the receptor neurons involved in psychological reactions are those responsible for our perception of pain and emotion. as well as interpretations of audio and visual stimuli. Psychoactive drugs. for the most part. alter or mimic the behaviour of four kinds of natural neurotransmitters: acetylcholine, norepinephrine, serotonin. and neuropeptides. The way in which THC works is still unclear.

Caffeine appears to act as a stimulant by activating intracellular metabolism,and alcohol is a general central nervous system depressant. Acetylchonine is released by neurons of the brain and the peripheral nervous system. It is a neutral transmitting chemicals. It causes muscle contractions and also speed up the heart beat. The compund is synthesized within the transmitting neuron and stored until the neuron is triggered. Once stimulated, the neuron releases acetylcholine into the synaptic region. After a neighbouring neuron has received the chemical at specific places on its dendrite and reacted, the acetylcholine is broken down by an enzyme so that it can no longer produce an effect.

Drugs can effect this sequence by blocking the transmission of acetylcholine (atropine and scopolamine), preventing the breakdown of the chemical, or by mimicking its action (nitotine). The drugs prevent the action of acetylcholine or neurotransmitter, prevent rapid muscle reaction and thus produce a relaxed sensation. Compounds that prevent acetylcholine breakdown, or mimic its effects, act as stimulants because they cause the receptor neurons to fire at an increased or continuous rate. Norepinephrine is another of the brain's endogenous (made within the body) neurotransmitters. Synthesis of norepinephrine takes place in transmitting neurons and released ofter receiving stimulus. However, after it exerts its effect, norepinephrine is reabsorbed, not broken down. Apparently, the same molecules are reused over and over. Some drugs (e.g., reserpine) deplete norepinephrine. Others such as cocaine prevent reabsorption of norepinephrine or mimic its action (mescaline, myristicine, and elemicin).

Serotonin is a brain neurotransmitter and it acts on the cells regulating body temperature, sensory perceptin, and sleep. Some psychoactive drugs such as the LSD-type compounds appear to alter the functioning of the neurons that transmit serotonin, thereby producing illusions of strange images. Finally, a recently didsovered group of neurotransmitter are peptides, polymers of amino acids. These include enkephalin, beta endorphin, and oxytocin. These chemicals are produced in minute quqntities and are received by very specific receptors. Some of these appear to act like the body's own pain killers. It is believed that opiates effect the same receptor sites as these endogenous pain killers and thus produce a dull, relaxed sensation. The evidence for this hypothesis isi that opiates are very specific in their effects, effective in low concentrations, and blocked by antagonistic agents. Opiates influence receptor cells in the brainstem, the medial thalamus, and at several places in the spinal cord.

History of Drug Use

We tend to think of our society as "the drug culture," but the use of mindaltering drugs is very ancient. What perhaps distinguishes our culture from previous ones is the modern dissociation of drugs from formal cultural or religious customs. It is easy to understand why humans would appreciate a substance that alleviated pain or reduced hunger and fagigue. At this stage the psychoactive properties of drugs merge into medicinal usage. Nevertheless, many drugs have always been taken in excessive quantities when ther was no physical need for them. These doese led to audio and visual hallucinations. The original appeal of these kinds of psychological effects was quite different from the modern enjoyment of the sensations as recreational experiences. Primitive peoples were unknown about hte natural phenomena and other events that was going around them and they used to call the action of supernatural agents.

Under normal conditions, people cannot see or speak with such powers, but under th einfluence of psychoactive agents, they could be transported to another world where communication with gods, demons, or even the dead was possible. In most cultures, only specific individuals were allowed to ingest psychedelic substances. Such people, usually men, were variously called healers, medicine men, witches, or shamans. Shaman, originally a Siberian word, has been adopted by anthropologists for this group of diviners because it lacks the connotations attached to most of the other designations. The initiation of an individual into the role of shaman was often painful and dangerous. Initiation rites sometimes involved starvation, some form of self-mutation, and consumption pf psychoactive drugs.

The combined effect produced a physical condition that enhanced the illusion of visiting the supernatural world and acquiring the insights necessary to serve as the intermediary between the earthly and spiritual worlds. The shaman took the drugs repeatedly whenever he needed the advice of ancestors or divine guidence to solve a problem. Anthropologists and botanists who have studied psychoactive drug use have found that New World peoples employed many more species of plants for their psychoactive properties than their counterparts in the Old World (40 versus 6 species).

It is assumed that the distribution of psychoactive drug plants is uniform in the two hemispheres, some authors have suggested that the explanation for the use of few plants in the Old World lies in the cultural superiority of Eurasian civilizations. This explanation assumes that shamanism is a primitive trait and that the cultural environments of the ancient European and Asian civilizations provided a milieu that replaced the need for spiritual intermediaries. Such authors forget, however, the localized distributions of many of the plants used by New World peoples and the overwhelming dominance of two psychoactive drug plants, the opium poppy and marijuana, in the Old World.

Cannabaceae

According to a Neolithic Chinese legend, the gods gave humans once plant to fulfill all needs. The plant was *Cannabis sativa*, ma, marijuana, or hemp. This assertion is not so far-fetched. *Cannabis* assuredly ranks among the world's most remarkable plants. It produces highly durable fibres that can be turned into ropes, fish nets, and clothing, its peeds are highly nutritious, and oil obtained from it, used in lamp or in paints and varnishes. Ten-thousand-year-old pot shards imprinted with twisted hempen fibers

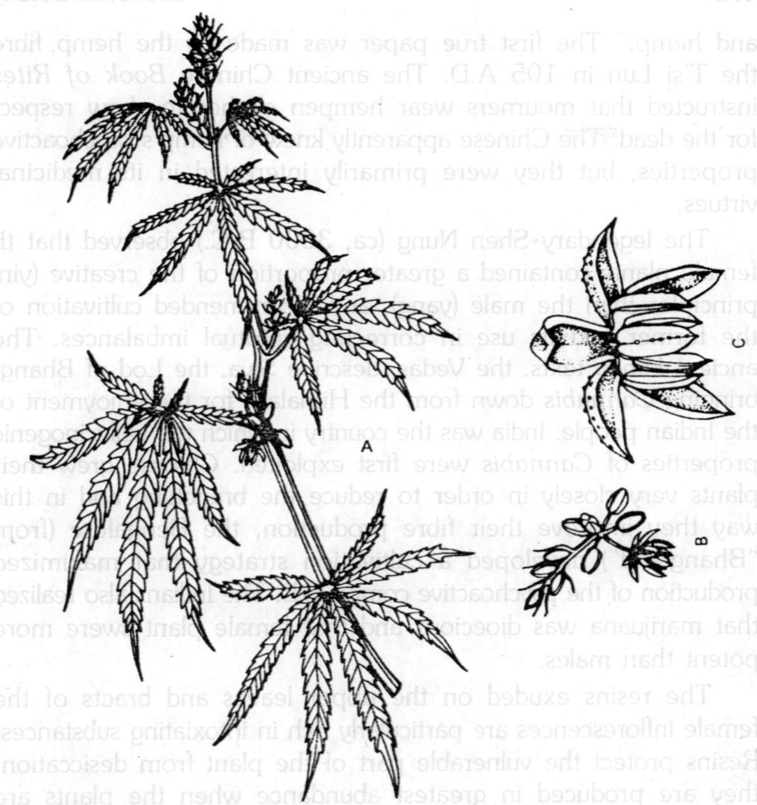

Fig. 12.1. Cannabis sativa. A—Male shoot; B and C—Flowers.

are part of the evidence that *Cannabis* was among the oldest cultivated plants. Today, the species is best known for the psychoactive chemicals it produces. Marijuana rivals alcohol, caffeine, and nicotine as the most widely used nonmedical drugs. It is said to be the most profitable of California's agricultural commodities and perhaps the second most lucrative crop in the United States. How it came to occupy this position is a fascinating story, involving cultures spanning almost every continent.

Cannabis, native to central Asia, the disperson of the seeds of *Cannabis* takes place by means of water, wind, birds or animals. Still, it is humans who have spread the species worldwide, often without knowledge of its psychoactive properties. The first to use *Cannabis* were the Chinese, who had such high respect for the plant that they referred to their country as the "land of mulberry and hemp." The first true paper was made by the hemp fibre the T'si Lun in 105 A.D. The ancient Chinese *Book of Rites* instructed that mourners wear hempen clothes to show respect for the dead. The Chinese apparently knew of hemp's psychoactive properties, but they were primarily interested in its medicinal virtues.

The legendary Shen Nung (ca, 3000 B.C.) observed that th female plants contained a greater proportion of the creative (yin) principles than the male (yang), and recommended cultivation of the former and its use in correcting spiritual imbalances. The ancient Hindu texts. the Vedas. describe Siva, the Lod of Bhang, bringing *Cannabis* down from the Himalays for the enjoyment of the Indian people. India was the country in which the hallucinogenic properties of *Cannabis* were first exploited. Chinese grew their plants very closely in order to reduce the branching and in this way they improve their fibre production, the Bėngalese (from "Bhangland") developed a cultivation strategy that maximized production of the psychoactive compounds. The Indians also realized that marijuana was dioecious and that female plant swere more potent than males.

The resins exuded on the upper leaves and bracts of the female inflorescences are particularly rich in intoxiating substances. Resins protect the vulnerable part of the plant from desiccation, they are produced in greatest abundance when the plants are exposed to heat and sun. By growing the plants widely spaced and removing the male plants (to prevent fertilization and prolong female flowering) resin producton is enhanced. The stimulation of resin production by dry conditions is the reason why plants grown in semiarid climates are more potent than those from cool temperate regions.

Likewise, sinsemilla (unfertilized, or without seeds) marijuana is one of the strongest forms. In India *Cannabis* is taken in the form of bhang, a milk-based beverage concocted with ground *Cannabis* leaves, sugar, and an assortment of spices. Bhang is widely drunk and is commonly offered as a gesture of hospitality. Indians classify *Cannabis* products into ganja, consisting of the potent female flowers and upper leaves, and hashish (charas), which is relatively pure resin. There are many colourful stories about traditional hashish collection. In Nepal, naked men used to run through the fields of flowering female plants and then scraped off the clinging resin globules. In Persia, the plants were beaten on rugs which wer subsequently washed to dislodge the resins.

Marijuana may have also had a role in Western civilizations, but it is generally believed that the species had not spread farther west than Turkey in ancien ttimes. The history of Arab use of hashish is not documented, but many legends exist to fill this gap. One of them credits an ascetic monk with finding Cannabis growing wild on the hillsides and sharing it with the poor (Sufis). The monk Haydar was so delighted with the plant that he ate nothing else until his death 10 years later in 1221 A.D. The Sufis used hashish to comfort the pain of their cheerless existence. When Marco Polo returned from his travels to the east in 1297 A.D., he brought back a story that was destined to become a classic. He recounted that there was a terrorist, Hasan-ibn-Sabah, the "Old Man of the Mountain," who had a large band of followers willing to do anything, even commit murder or sucide, for their leader. This blind loyalty was attributed to a very clever recruitment strategy. Political candidates were drugged and brought unconscious into an exquisite mountain garden. There, among the exotic flowers and water works, beautiful women were ready to minister to every need.

After experiencing the pleasures of this paradise, the recruit was drugged once again and, upon awakening, told that he could return to the garden in this life or after death only if he swore complete allegiance to the Old Man. The technique was apparently very successful, and the band of loyal followers became known as assassins, a name purportedly derived from hashish, the drug they were supposed to have been given.

There is no evidence to substantiate this story, and much to refute it, but it was compelling enough to be retold countless times and has been misinterpreted even in recent history by antimarijuana zealots who imply that the hashish produced violence rather than to ensure loyalty to the assassins. In the fifteenth and sixteenth centuries, Arab traders introduced marijuana to Africa where its use quickly spread. The drug was commonly given to women in childbirth and fed to babies when they were weaned. The Kafirs called the plant daga, a name later applied to a culture that made widespread use of *Cannabis*. The dried plant part was mixed into beverages which was taken by Africans. The technique of smoking became popular only after the Dutch colonized the continent. By the time of the Crusades, when European contact with the Arabs was reestablished, marijuana use was common throughout Africa and Asia.

The *Tales of the Arabian Nights* (written sometime between 1000 and 1700 A.D.) was widely read in Europe and provided a romantic introduction for many to hashish. "The Tale of the Hashish Eater," in particular, described the effects of the drug ona degenerate who used it to delude himself that he was wealthy and handsome. When Napoleon's army was stationed in Egypt in 1798, recrutis experimented with the local intoxicant and returned to France with stories and samples. The French student visited the northern Africa and returned back with glowing youthful reports of marijuana use. A frecnch psychiarist, Dr. Jacques Joseph Moreau de Tours, read the medicinal uses of hashish in combinating plague and dysentery and decided to test the drug himself. Because he had read that the drug produced drastic change in perception, he thought that he might be able to administer it to sane individuals in order to reproduce the symptoms of psychosis and thereby learn about the causes of mental illness. He asked an artist friend, Theophilie Gautier, to try the experiment of the drug and report his reaction.

Gautier was impressed with the effect of hashish on hsi artistic senses, and he recommended it to his frient, the painter Boissard. Eventually the group expanded, meeting monthly in Boissard's hotel suite where they consumed hashish while being observed by Dr. Moreau. The 'Club des Hachichins' named after the story of the Old Man of the mountains by themselves. The club eventually became a meeting place of the great French artists and writers of the day: Alexander Dumas, Victor Hugo, Eugene Delacroix, and Charles Baudelaire.

Not surprisingly, much of the art and literature of this period reflects the preoccupation with the newly discovered drug. Despite Moreau's eventual conclusions published in "On Hashish and Mental Alienation" (1845) that hashish use had overall deleterious effects on mental health, the use of the drug continued to spread. The fascination with hashish travelled across the English Channel, from the French to the British intellectual community, where W.B. Yeats, Oscar Wilde, and Ernest Dawson experimented with it to see if it enhanced creativity. After introduction of Marijuana into the Americans it was scattered all over the country The first introduction of cannabis to the New World was done by the Spaniards during hemp cultivation in Chile in 1545.

The British, hoping to produce a lucrative attempts failed, *Cannabis* managed to escape cultivation and spread around the island. When African slaves were brought to Jamaica in the middle of the century to harvest sugar cane, they found a local source of their familiar drug already established. Before 1800, the British had sent formal orders for the colonists in North America to produce hemp. Even Thomas Jafferson and George Washington realized the military need for rope-making fibers and started hemp cultivation. An appreciation of the other uses of *Cannabis* did not develop in the United States unti the 1800s. The introduction of hemp's psychedelic effects followed the same pattern as in France and England.

First there was tremendous curiosity, primarily by intellecturals and artists, a rash of drug-inspired art, and then a general spread of marijuana use. In fact, the 1876 Centennial Exposition had a Turkish bazaar which featured hashish smoking as a special attraction. Despite the sensatinalized stories that soon began to appear about decadent hashish dens and violent consumption by the poor in such manner the use of the drug spread. During prohibition, marijuana smoke floated up and down the Mississippi to the tunes of Dixieland. Jazz musicians found "moota" a pleasurable way to enhance their music without experiencing the stupefying effects of alcohol. They expressed their appreciation of the drug with such songs as Benny Goodman's "Sweet Mrijuana Brown," Cab Calloway's "That Funny Feefer Man," "Texas Tea Party," the "Mary Jane Polka," and others. It has been suggested that the desire to suppress the increasing popularity of black and Latin-inspired music began the movement that eventually led to the criminalization of marijuana. However, racism would have been only one of many factors that countered marijuans use.

Harry Anslinger was trying to justify the existence of the newly created Narcotics Bureau of the Government. He also carried out the most effective antimarijuana campaign. He publicized stories of the horrors of marijuana "addiction" and depicted the put smoker as a savage fiend whose aroused sexual desires and violent tendencies led to criminal activities and the use of stronger drugs. The story of the "Old Man of Mountain" was revived by Anslinger but the story was twisted a little so that the drug iteself was the cause of irrational violence. Marijuana was tried by the press and condemned by an emotional, ill-informed public. In the late 1920s and 1930s, marijuana use was banned in Louisiana, Texas, and Illinois.

Other states soon followed, and in 1937 the use of cannabis came under federal jurisdiction withe the passage of the 1937 Marijuana Tax Law which actually heavily taxed, but did not prohibit, the drug. Other countries had previously taken hard step to prevent Cannabis use. In this field South Africa passed the first anti-Cannabis law in 1870, and by 1925, the League of Nations agreed to an international statute against its use. All of these legal sanctions against *Cannabis* had two things in common, they were not based on any substantial medical evidence of severe effects of marijuana use, and they universally failed to halt its spread. Most medical studies beginning in 1893 with the Indian Commission's through report failed to find any detrimental effects of moderate use. Sanctions against marijuana only seem to increase its popularity.

In the 1960s the explosive spread of pot smoking provides the drastic demonstration of such kind of effect. Young people throughout the world, disillusioned with the ways in which the older generation was handling affairs, made marijuana smoking a symbol of their rebellion. While attention was focused on marijuana in the 1960s, the suggested medical virtues as well as the possible detrimental effects were investigated. Once THC was finally isolated in 1965 and measured quantities could be used for testing, it was discovered that the active principal is effective in reducing the pressure exerted against the eyes of glaucoma patients. Cancer patients undergoing radiation or chemotherapy treatment when used such compound, the nausea problem reduced. Since THC dilates the bronchial vessels, it provides relief for asthma sufferers. Because of the useful medicinal qualities, marijuana cigarettes or concentrated THC in the form of pills can now be prescribed by physicians when appropriate. In 1972, a federal government study on the effects of marijuana was released. The report suggested decriminalization of marijuana because there was no evidence of medical or mentla damage from use of the drug and because the drug was so widely used that laws against it were unforceable.

In 1981, the results of another study made by the National Academy of Sciences were released. The academy panel observed the use of marijuana very carefully and came to conclusion that the low or moderate use weakened the sense, sensibility and sensitivity of users. They could find no evidence of addiction or permanent deleterious medical effects with low or moderate use, but heavy use was correlated with several psychological and physiological problems. The use of marijuana is directly linked with low sperm count in males. Since marijuana is usually smoked, heavy and regular user suffers lung disorders similar to those incurred by cigarette smoking. Finally, the Academy found that ther was a high probability that heavy users would turn to "hard drugs." The panel did conclude, however, that marijuana was safer thanits legal, more widespread counterparts, alcohol and tobacco, and suggested that the present legal sanctions against possession and trade be removed.

Papaveraceae

In United States poppies are well known as garden ornamentals or for their culinary seeds, but they rose to fame because of opium. The capsules of the fruits from which the seeds are obtained are rich in an alkaloid-containing latex. The latex has medicinal use but opium and morphine these are the most abundant of the opium alkaloids and are used as mind-altering drugs. The collection of opium from fruits of the opium poppy (*Papaver somniferum*) extends back to at least 3000 B.C. Sumerian tablets from 2500 B.C. refer to opium as the "joy plant" and describe the ingestion of smal balls of the latex to induce sleep and relieve pain. Opium is mentioned in every medical treatise written during ancient Greek and Roman times. The name opium comes from the Greek "opion" for poppy "juice."

The early Greeks realized that the driniking of wine in which latex had been dissolved led to trancelike state, and they therefore associated poppies with several divinities such as Hypnos, the God of Sleep, Morpheus, the God of Dreams, and Thanatos, the God of Death. It is commonly known that Chinese was responsible for introducing opium to the rest of the world, but China was ignorant of the drug until the seventh century A.D. when Arab traders first brought samples to the Orient. In the East, it is used to cure dysentory in some manner that paragoric (tincture of opium) is used for diarrhea today.

In the seventeenth century, the Dutch introduced tobacco smoking to Formosa and began to mix tobacco with opium in their pipes as a treatment for malaria. The practice of smoking opium spread to the mainland where it was rapidly adopted throughout the country. Soon, tobacco disappeared from the mixture, and smokers inhaled vapours from heated balls of latex that were droped into the bowl of a pipe. Opium smoking became so popular that Chinese officials thaught to ban the sale of opium. Their antiopium edicts, issued as early as 1729, were ignored by both the Chinese populace and th Portuguese suppliers. In 1800, the British gained a monopoly over trade rights with China. Since China had little use for of opium dealing because it was one of the few commodities for which the Chinese would exchange coveted products such as silk and spices. Eventually, England even established opium plantations in her Indian colonies to provide opium for trade.

Importation of opium into England was, however, illegal at this time, but distance and the need to maintain international trade seems to have exonerated the practice elsewhere. The Chinese lords tried again and again to halt the drug trafficking because they could see the debilitating effects on their people, but without the support of the British, their efforts failed. The situation came to a head in 1839 when the Chinese confiscated and destroyed British opium supplies in Canton. England retaliated by invading China, touching off the first Opium War. The war ended wiht a treaty that ceded Hong Kong to the British.

A second Opium War broke out in 1865 and the Chinese were again defeated. This time they were forced to agree to British opium importation. Naturally, opium consumption among the Chinese continued to escalate. Trafficking in the drug was finally reduced in 1906 when the last emperor of China managed to get the British to agree to a reduction in importation. Still, use did not drastically drop until the revolution and the establishment of the People's Republic of China in 1949. Opium trading and use are now strictly forbidden. Opium consumption was not common in Europe until 1525 when Paracelsus discovered, or rediscovered, a way to dissolve it inalcohol.

The opium is dissolved in alcohol to obtained the tincture was known as laudanum. At that time laudanum was the most popular drug due to its meaningful medical benefits of the case with which it could be consumed. In 1803, morphine was isolated form opium, producing a purified alkaloid that could be given in measured doses. On a per weight basis, morphine is 10 times as strong as opium (which contains at least 24 other alkaloids andnumerous other compounds). Purified morphin can be administred intravenously to releaf the pain. When the hypodermic syringe was developed in 1853, it presented doctors with a rapid method of introducing this potent pain killer into the bloodstream and morphine use rose. Morphine's effectiveness was both a blessing and a curse. In many cases, the speed with which a drug enters the bloodstream (and thus the intensity of the "rush") is correlated with the level of addiction it can produce.

Only after 45,000 soldiers returned home from the Civil War addicted to thepain killer were the dangers of morphine use recognized. In an attempt to develop a nonaddicting pain killer, scientiets discovered that morphine could be chemically altered by the addiction of two acetyl groups. The end product is a semisynthetic compound known as *heroin*. Heroin is an even more powerful analgesic than morphine, and it, like morphine, was initially explained that heroin contain less addicting qualities of its parent drug. While mark of this idea persisted untill 1905, but very soon become apparent that heroin was even more dangerous than morphine because it crosses the cell membranes very easily and rapidly than its natural counterpart. Because of this fast absorption (and hence cycling around the bloodstream), heroin must be more frequently administed than other opiates. In addition to the effects produced by heroin itself, users of the drug suffer from disruption of the blood flow, infections, diseases such as hepatitis, and collapsed blood veins.

Heroin is very physically addicting and produces pronounced withdrwal symptoms once the habit has become established. Two to three weeks of injecting 30 mg per day is considered enough to create a tenacious habit. The main cause of death for heroin addicts is by overdose. Because of the common usage of adulterantnts in heroin and the use of heroin in combination with other drugs, "overdose" levels for an individual ar enot constant and therfore cannot be adequately gauged. The lifestyle of the heroin addict who no longer seeks the drug for pleasure, but because he or she desperately needs it to avoid withdrawal, tends eventually to lead to overall physical problems. The death rate for heroin users is more than twice the normal rate. In 1914, a law was passed prohibiting the possession of opiates for nonmedical purposes. In 1924, the production of heroin was proved illegal in the United States and in 1956 existing legal supplies of the drug were destroyed completely. Because of its addictive properties, even morphine is seldom used in its pure form as a modern-day medicine.

In contrat, the use of heroin, all illegal, has dramatically increased during the last 30 years. Since heroin is made from morphine, the total production of opium has correspondingly increased. In 1972 most of the opium production ended up in the United States but still grow in Turkey. When Turkey and the U.S. government reached an agreement to limit opium poppy cultivation, th ecenter of productio shifted to southeast Asia. Economically, opium is an extremely lucrative crop. In 1972 10 kg of crude opium which sold for $250 where it was grown, would eventually gross $200,000 when sold on the street. With profits such as these, people are willing to risk the penalties if caught.

Fabaceae

A large number of new wrold Indian tribes used the species of the bean family which is rich in alkaloids, as hallucinogenic agents. Snuff made from powdered pods of cohoba (*Anadenanthera peregrina*) was first reported form Hispaniola in 1496 and was much later found to be used by the Indians of the Orinoco region of Venezuela. In South America, the snuff is called "yopo," and it is often mixed with lime (Calcium carbonate) when taken.

The snuffs can be inhaled througha number of devices, one of which is a long, hollow pole. One end of the pole is placed in a nostril and the other inot a pile of snuff on the ground. Mescal, or red bean (*Sophora secundiflora*), a common ornamental shrub or small tree of the American southwest, was used by native proples of this region to induce trances. The plant has received some recent publicity because mescal beans have been confused with mescaline and unknowing youths have been tempted to try them. However, *Sophora* contains cytisine, a much more dangerous alkaloid than mescaline, and one that can easily cause death due to respiratory failure if slight overdoses are taken.

Celastraceae

As Americans, we are accustomed to take a cup of coffee, tea, or other caffeine-containing beverage when we need a lift or freshness, but in other parts of the world the same effect is

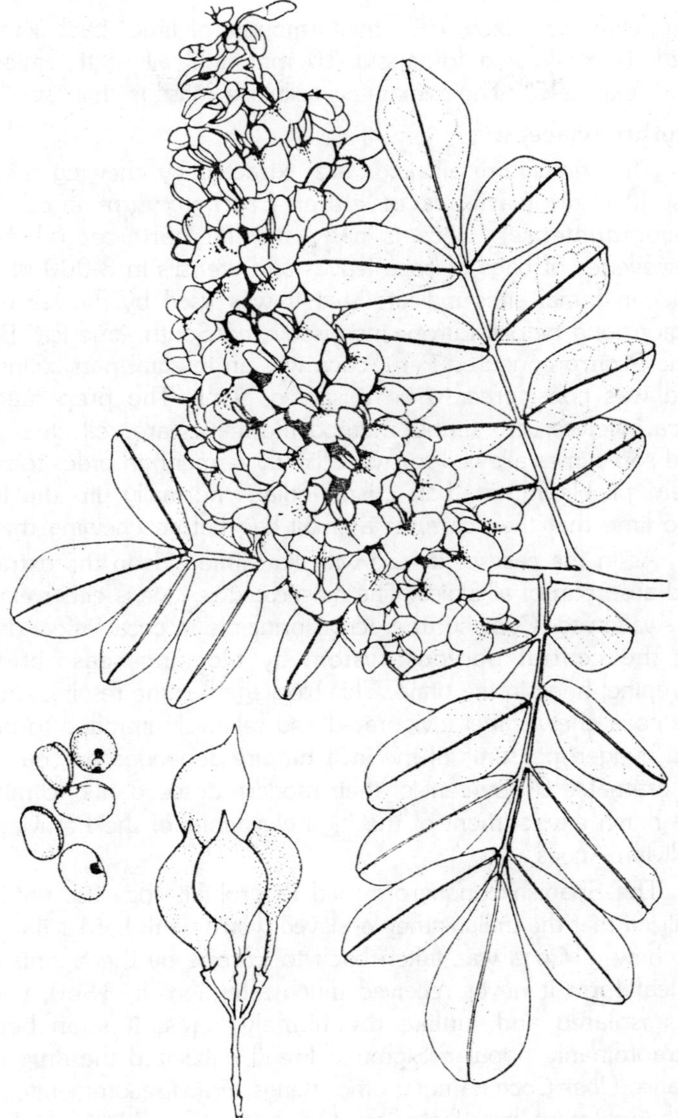

Fig. 12.2. Sophora secundiflora is valued as an ornamental plant because of its fragrant purple flowers and glossy green foliage.

produced by consuming parts with different but equally effective stimulatory alkaloids. Among these is kat (*Catha edulis*), a native of Arabia but now most widely grown along the eastern edge of Africa. The alkaloids are ingested by chewing wads of freshly cut leaves usually mixed with small amounts of lime. Each lump, or quid, is masticated for about 10 min until all of the juice has been expressed. The remaining cellulose mass is than swalloed.

Erythroxylaceae

The stimulatory alkaloids are extracted by chewing the coca just like kat (varieties of either *Erythroxylum coca* or *E. novogranatense*). Coca is native to the north-central Andes. Discoveries of bags of coca leaves and utensils in 3,000 year old Andean burial sites indicate that it was used by natives of this region long before Europeans discovered South America. By the time Pizarro conqured Peru, coca was an integral part of Inca life and was considered to be a sacred plant. The preparation of coca involves very simple method the leaves are collected, dryed and sometimes allowed to sweet (lightly ferment) in order to render them pliable rather than crisp. Indians normally dip the leaves into lime that they carry in a small bag before chewing them.

As in the case of its use with kat, lime aids in the extraction and absorptin of alkaloids. The masticated residue is either expelled or swallowed. Cocaine, the active principle in coca, affect directly on the central nervous system by blocking reasorbtion of norepinephrine in the brain. This blockage has the result of making the coca chewer feel invigorated and relatively immune to fatigue and hunger, peasants of the Inca Empire depended on coa, as do an estimated 90 percent of their modern descendants, to mitigate the harsh environment of the high elevations of the Peruvian and Bolivian Andes.

The Spanish conquerors tried to prohibit coca use until they realized that the Indians they enslaved would work harder if allowed to chew it. Coca was taken back to Europe by the Spanish, but in leaf form it never received much attention. In 1860, cocaine was isolated and, unlike the homely leaes, it soon became extraordinarily popular. Sigmund Freud publicized the drug in his treatise *Uber Coca*. Among other things, Freud recommended coca as a treatment for alcoholism and morphine addiction. He also lauded it as a local anesthetic and praised its use in psychotherapy to relieve depression. In both Europe and the United States, doctors began to prescribe its use.

An enterprising Italian, Angelo Mariani, developed a coca wine beverage in 1860 which was soon the rage of Europe. Mariani's wine earned commendations from scores of celebrities including

Jules Verne, Ulysses S. Grant, President McKinley, Emile Zola, Henrik Ibsen, and Thomas Edison. Coca cola; which originally consist both caffine-rich extracts from Cola nitida and coca txtracts, was first marketed in 1880 as a headache remedy. Since 1904, however, federal law has prohibited the inclusion of cocaine in nay beverage. Ironically enough, after the Coca Cola Company complied with the federal law, it was sued for misleading advertising on the groudns that the name implied that the beverage contained coca products.

As a result, coca leaves, with the cocaine removed, are now used to flavour the syrup from which the soda is made. As more and more reports of cocaine-induced violence and other unsubstantiated effects of the drug circulated after the turn of the century, the government took steps to curb its use. In 1914, the drug was formally declared illegal by the Harrison Narcotic Act. As a result, cocaine costs escalated, but despite illegality and high costs, its use continues to increase. Status seekers appear delighted to pay exorbitant prices for a brief experience. Indulgers usually chop or grind crystals of cocaine hydrochloride before sniffing ("snorting") a line of the fine powder. Cocaine may also be smoked or injected. The conversion of cocaine from the salt form to the base form is termed 'Free basing', which alters its solubility (increasing the drug's potency), has proved to be particularly more harmful. Both these latter methods of ingestion produce stronger sensations but are much more dangerous to the user than sniffing. Just how dangerous is cocaine? While researchers have not been able to find any organic damage from prolonged coca use per se among Andean Indians, their findings do nto mean that cocaine is a harmless drug.

Indians used to chew coca leaves and so they absorbs only limited quantities of the drug because the leaves contains alkoloides in relatively small amount (0.65 to 1.25 percent on a dry weight basis). An average native user might consumes 2 oz of leaves per day or about 0.7 grains of cocaine. A person who "snorts" or injects cocaine could be consuming as much as 6–8 grains per day of the drug plus adulterants. 1.2 grams of cocaine is considered as a lethal dose. Cocaine is not physically addiciting, but its use can become a pronounced habit as indicated by the burgeoning number of treatment centers for cocaine abusers.

After becoming accustomed to the elevated feelings it produces, habitual users suffer from depression when denied the drug. While occasional use produces feelings of mood elevation, vitality, mental clarity, sexual stimulation, and reduction in appetite in many users, chronic users can suffer from anxiety, conofusion, insomnia, and impotence. Paranoid psychosis can develop from longterm heavy use. Cocaine ingestion of pregnant animals is harmful to fetal development. This is concluded by recent experiments. As more reserach is conducted, other harmful effects may be discovered.

Malpighiaceae

One of the most important groups of plants used by South American Amazonian tribes as a source of hallucinogenic compounds are members of teh genus *Banisteriopsis*, primarily *B. caapi*. Depending on the tribe, this species is called caapi, ayahuasca, yaje, or cipo. Infusions of mashed bark or stems are usually prepared, but stems can also simply be chewed in order to release the active alkaloid. People using ayahuasca describe the perceptions produced as feelings of having experienced death or of experiencing a separation of spirit and body. Sometimes

Fig. 12.3. Banisteriopsis caapi is a tropical vine with stems that contain the alkaloid hormine, a psychoactive substance known as yaje or ayahuasca by the Amazonian Indians who use it in rituals.

the hallucinations are pleasant, other times apparently terrifying with vivid apparitions of jaguars or snakes. Like many other alkaloid-rich preparations, mixtures of ayahuasca often cause vomiting and diarrhea. Unlike the use of many other psychoactive drugs by native peoples, ayahuasca is often taken in communal rituals.

It is thought that this practice of communal ingestion enhances feelings of clairovoyance or telepathy. The group, under the supervision of a yaquero, is led through chants to experience shared hallucinations. This effect was reflected in the name "telepathine" given to the mixture of active chemicals first isolated from *Banisteriopsis caapi*. Since this initial chemical work, the active compounds have been further purified and named harmine and harmaline. In some tribes, the rituals commemorae an act of incest by a deity. In almost all cases, the ceremonies are filled with sexual symbolism, but there is no sexual gratification during the experience.

Cactaceae

A large number of species of the Cactaces family are good source of hallucinogenic compounds. The best known is peyote, *Lophophora williamsii*, a small, globose, gray-green cactus native to the Rio Grande Valley bordering the United States and Mexico. Native American Indians appear to have extended its distribution to Arizona and New Mexico centuries ago. No one is sure when peyote was first used, but sixteenth century reports by European explorers describe its use by the Aztecs as a divinatory plant.

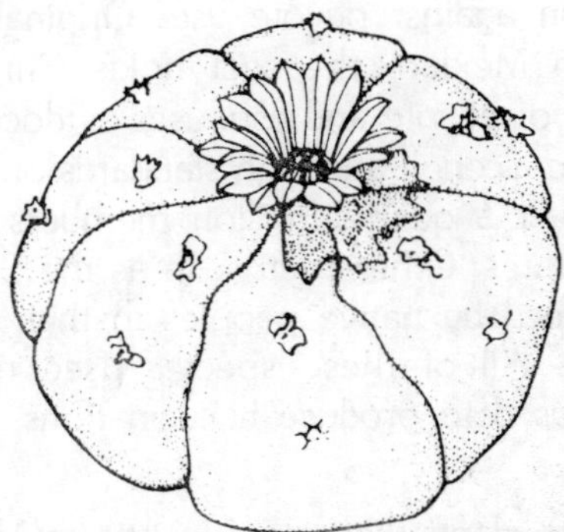

Fig. 12.4. Mescaline is the psychoactive compound in the gray-green peyote buttons that grow natively in southwestern U.S. and adjacent Mexico.

These accounts refer to peyote as the "diabolic root," because the Spanish observed the Aztecs using the plants ritualistically.

After the collapse of the Aztec Empire, use of peyote survived among a few Mexican Indian tries such as the Huichol, Cora, and Taramara. Inthe United States, the Plains Indians began to use it as late as the 1880s. In India harvesting is done by cutting off the top of the stems and leavign the sturdy taproot for regeneration. The stem tips, or buttons,were eaten fresh, or dried for later consumption. The dried buttons can be swallowed by softening and mastication. The initial reaction after swallowing in nausea. One to two hours later, quesiness disappears and is replaced by kaleidoscopic illusions of vivid colours, sensual hallucinations, and distorted perceptions of time. During this last period, which can last from 5 to 12 hours, the faithful report hearing the voices of their ancestors who help them to diagnose and cure any problems they may have.

Consumption of peyote means ingestion of 30 to 40 different alkaloids and several chemicals at a time, mescaline is one of the most active compound present in the alkaloids. Isolated in 1896, mescaline was the basis for research into psychedelic compounds for the next 50 years. While there is no evidence that peyote or pure mescaline is addicting, repeatrd use may permanently affect the mind. Because of their potentially harmful effects, both the plnat and the compound are illegal to possess or sell in the United States. However, one religious sect, the Native American Church, uses peyote as an integral part of its services. In a case that went before the Supreme Court, the church received an exemption from the Prohibition against peyote use. Originally founded by natives in northern Mexico, the sect holds Christian services modified by the use of peyote and espouses as doctrine, brotherly love, abstinence from alcohol, and high standards of moral conduct. Today, there are over a quarter million members of the church within the United States. Other cacti such as the San Pedro cacti have been widely used by native peoples in their native regions of Peru and Chile. All of these species used mostly contain mescaline and are used to produce hallucinations.

Apocynaceae

While fewer drug plants were used by native Africans than by South and Central American natives, one, iboga (*Tabernathe iboga*) was important as an hallucinogen. Iboga is, in fact, one of the very few psychoactive drugs in the entire periwinkle family. Although th species of the family are rich in alkaloids, most of them are toxic. A few species such as *Rauwolfia serpentina* and *Catharanthus roseus* are used medicinally, but there is no indication that they were used in rituals. In preparing iboga, the bark or roots of the shrubs are powdered or chewed, often with other plant species. Ingestion of the alkaloids, one of which is ibogamine, is said to lead to frightening visions and in large doses can produce severe convulsions, respiratoty arrest, and death.

Convolvulaceae

No species of morning glory family were known by scientists before 1903, which could be used for psychoactive purposes. In 1955, Humphrey Osmond described the use of *Rivea corymbosa* by Mexican natives, and, in 1960. T. MacDougall detailed the use of another member of the family, *Ipomoea tricolor* for altering perceptions. Still more startling was the discovery in 1960 that the sees of these plants contained D-lysergic acid amide. Prior to this time, lysergic acid alkaloids were thought to be restricted in nature to ergot (*Claviceps purpurea*), a rust that infests grains, and to a few other fungi. Synthetically produced LSD had become

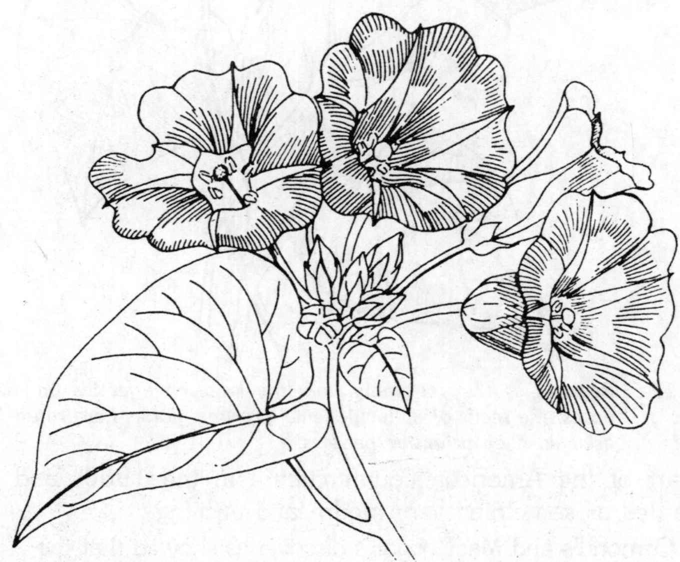

Fig. 12.5. Rivea corymbosa, or ololiuqui a white-flowered vine of the morning glory family contains the psychoactive compound, D-lysergic acid amide.

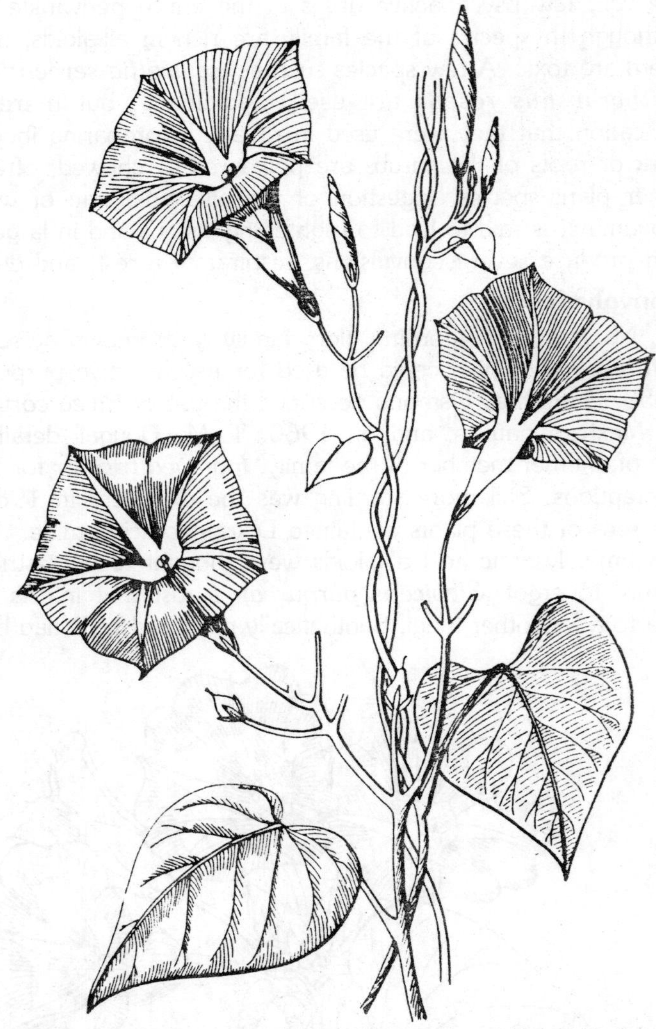

Fig. 12.6. Because of the extremely fine line between effective and lethal doses, the seeds of hallucinogenic morning glories have rarely been used even by primitive peoples.

a part of the Americna counterculture in the 1960s and was regarded as something very modrn and exciting.

Osmond's and MacDougall's discoveries showed that the effects of these kinds of compounds had been recognized and utilized by native peoples, perhaps for thousands of years. Both of the convolvulaceous species now known to be used are called *ololiqui* (pronounced o-low-lee-oo-key) by the peoples that use them. The carefully measured quantities of seeds should be ingested especially by experienced individual because there is a little difference between the quantities seeds that produced vivid hallucinations and the quantities that produce death, there has been little lay experimentation with the seeds, despite the fact that one of the species is a commonly cultivated ornamental plant.

Solanaceae

The plants of the Solanaceae is highly toxic therefore their use is restricted outside of some Central and South American people rarely used today as source of mind-altering drugs. Undoubtedly, the major reason for their restricted use is their toxicity. At one time, however, various species were used in both Eurasia and the Americas as sources of hallucinogenic compounds. Historically, the most important of these were species of *Datura*,

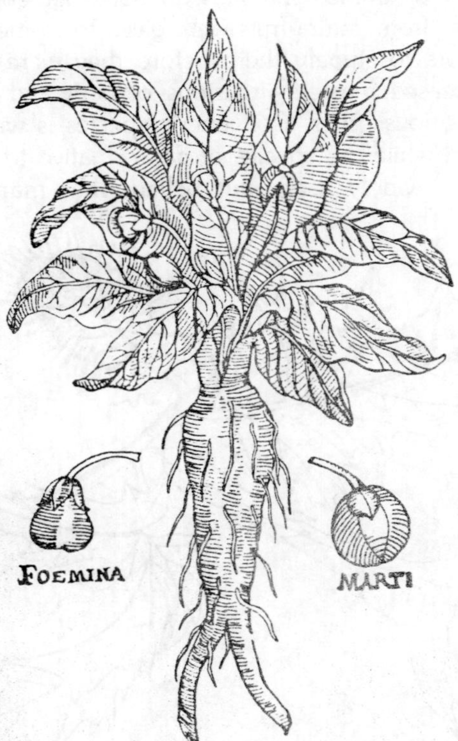

Fig. 12.7. Mandrake, or Mandragora was long thought to be an aphrodisiac because of the human shape of its root.

Hyoscyamus, *Atropa belladonna*, and *Mandragora officinarum*. Many of these were also used medicinally, and a few have become important modern sources of pharmaceutical compounds. The tropane alkaloids (atropine, scopolamine, and hyoscyamine) which these species contain are effective in controlling smooth muscle action. It is careful measured amount is useful in the treatment of heart irregularities. In larger, but sublethal, doses, the same alkaloids produce hallucinations. Of the plants in this group, only datura is native to the New World.

Datura is particularly rich in scopolamine, which is considered to be the most hallucinogenic of the solanaceous alkaloids. Seeds or roots of various species provides vision upon eating. So it is widely used in Central and South America, but even among people that use datura, it is considered dangerous and too strong for general use. Usually datura was ingested only by boys during puberty rites or by trained shamans. In South America, alcoholic beverages made from datura fruts were given to women and slaves of dead warriors to stupefy them before they were buried alive with their deceased masters. Infusions of bark and leaves were also used by various tribes. Congeneric species is widely used in the New World while the species of datura native to Europe and Asia were less widely used, but belladonna, mandrake, and

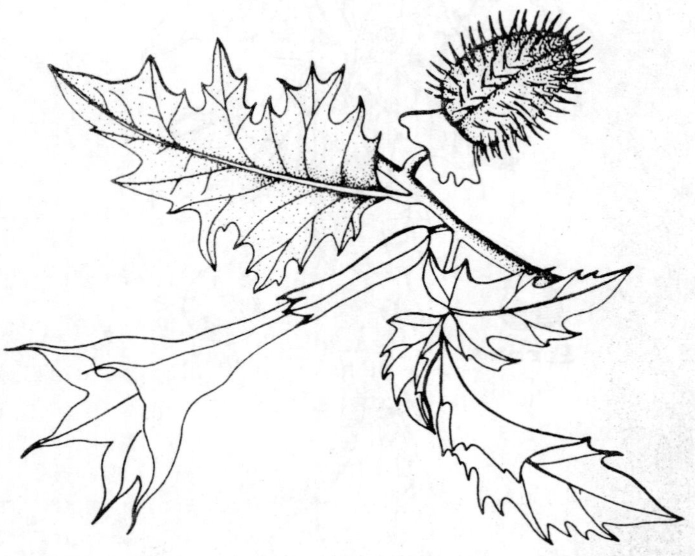

Fig. 12.8. Datura acquired its common name jimsonweed in 1705.

henbane assumed great importance in Europe during the Middle Ages.

The ancient use of solanaceous alkaloids in witchcraft and folk medicine led European scientists after the Renaissance to investigate their actionand eventually to the discovery of their medicinal properties. Likewise, their uses as hallucinogens and poisons prompted writers and artists to include them in the writtings and paintings of the time. Belladonna or henbane are the source of Atropine. Atropine is directly absorbed by skin, and during the Middle Ages, witches rubbed ointments containing it on their bodies. The psychological sensations produced give one the feeling of flying. While under the influence of atropine, witches were supposed to be transported to rendezvous with spirits of demons. These meetings, called sabats, are often figured in European art of the period. Our modern portrayal of a witch riding a broom may stem from these ancient rites.

The depiction of the flying comes from the hallucinogenic sensations of being transported, and the broom purportedly represents the stick used to apply atropine-impregnated ointements to vaginal membranes for absorption. Another figure commonly attributed to visions seen while under the influence of solanaceous alkaloids is the werewolf.Similar to the visions of jaguars and snakes produced when South American natives use caapi, wolves were associated with trances induced by henbane, belladonna, and related species. A quite different member of the Solanaceae, however, has had the greates tworldwide impact. N. rustica is believed to be the relative of Nicotiana tabacum. Both of them become important commercial items in almost every country of the world. The genus *Nicotiana* is thought to be native to the New World, but at least one species, *N. suavolens*, which occurs in Australia, was independently domesticated.

At least 1,000 years before Columbus landed in the West Indies, tobacco was smoked, eaten, and snuffed by native peoples throughout the New World. Among American Indians, tobacco was used medicinally to ease childbirth pains and stave off hunger on long hunts. The dried leaves of tobacco become so valuable that it becomes the source of money and incorporated into ritualistic and symbolistic practices such as the smoking of a peacepipe. Tobacco was considered sacred by many American tribes. Mayan priests thought that smoke rising from their pipes carried messages to the gods. Recent studies of Amazonian tribes have shown that the use of tobacco in divination rites continues today.

Shamans of the Warao tribe of Venezuela, for example, fast almost to starvation during their initiation rites and then eat and smoke large quantities of tobacco. The resultant hallucinations produce the sensation of being transported to another world for mettings with the spirits that govern the lives of the people. Later in their careers, shamans frequently turn to tobacco to help them solve tribal problems. Although Columbus brought "cigars" to Queen Isabella after his first voyage, it was not until Andre Thevet brought seeds of Nicotiana tabbacum from Brazil to France in 1556 that tobacco cultivation started in Europe. Linnaeus named the genus after Jean Nicot, the French ambassador to Portugal who made a fortunate importing and popularizing the use of the plant in Paris.

Tobacco spread across Europe and Asia and gained a special place because its medicinal virtues as a remed for several female problems, a snakebite antidote, lung strengthener ulcer remedy, cure for the plague and potent aphrodiasiae. Tobacco use became so widespread that King James I was alarmed to find that purchases of tobacco were depleting England's silve supply. Neither the high cost of the leaves nor King James's blistering attack in his *Counterblaste to Tobacco* (1604) checked the spread of smokingor sniffing. In other countries as well, leader tried to discourage tobacco use. So many hard step and stricked rule were formed in order to prohibit the smoking as one Chinese emperor ordered smokers to be decapitated and Russian Tsar ordered the nostrils of snuffers to be split so that they could no longer practice the habit. Still, tobacco continued to attract followers. Eventually, the British promoted tobacco cultivation in their colonies to ensure a national supply. Virginians started growing tobacco in 1612. Since an acre planted in tobacco yielded four times the revenue of an acre planted to corn, colonists turned more and more exclusively to tobacco farming. The Virginia monopoly on tobacco production granted by the Queen was broken when settlers in Maryland began to cultivate the crop.

Soon, much of the eastern seaborad of the United States was engated in tobacco production. For many years, tobacco was the most important item of trade between England and North America. Primitive method of cultivation of tobacco consisted simply of saving seeds of the previous year's crop and sowing them in a cleared area. After maturity plants are harvested and the leaves dried in the sun. In contrast, modern tobacco farming is quite elaborate and employs various cultivation techniques for the many different cultivars and the types of of tobacco being produced. On modern U.S. farms, tobacco is usually first sown in seed beds and the seedlings transplanted into fields after 2 months of growth. The spacing of plants and fertilizer regime varies with the final type of tobacco to be produced.

Nitrogen-rich fertilizers are applied to encourage the development of large, pliable leaves suitable for cigar wrappers, while the small, more brittle leaves favoured for cigarettes are produced by limiting nitrogen and supplyign phosphorus and potassium at critical stages. While it is growing, the flower stalks and the side shoots are usually removed to prevent the plants from diverting resources to seeds and low branch development. Leaves used for cigar wrappers should be large so the plants are generally grown under cheesecloth where the high humidity and protection from sunscald also promote formation of large, thin, blemish-free leaves. Two to 4 months after planting, the crop is considered mature Good grades can be picked a leaf at a time, or whole stalks can be harvested. In either case, the leaves or stalks are tied into bunches called "hogsheads" for curing. During the curing process, themoisture content in the leaves is reduced from about 80 to 20 percent. Starches are converted to sugars, and some proteins are broken down enzymatically.

Slow drying permits aerobic fermentation to take place but prevents the growht of molds or fungi. Curing can be done by circulating air or by smoking. Occasinally, bundles of leaves are sun-dried. Following curing, the leaves are aged for periods varying between a few months and a year. The tobacco used for the cigarette and cigar filling is remoistered before marketing and the veins and petioles removed. The softened leaves are then cut by machines into strips. Depending on the ultimate tobacco product, are then cut by machines into strips. Depending on the ultimate tobacco product, various materials can be added.

Mositure-retaining substances such as glycerin, cider concentrate, and diethyl glycol are common additives, but honey or sugar, oil of hops, licorice, coumarin, rum, or menthol can be added for flavouring. Additives have been increasingly used in tobacco mixtures since the middle 1970s to compensate for the flavour and aroma lost when manufacturers produce cigarettes with low tar and nicotine. While most of these are drawn from the FDA list of "Additives Generally Recognized as Safe," there is concern that, when burned, many may constitue health hazards. Coumarin, a compound used as additives removed from the FDA list of safe drugs because it has been shown to be carcinogenic, is stull being purchased in large quantities by cigarette manufacturers.

Locorice, likewise, produces carcinogenic compounds when burned. Even the sugars added produce tars when burned. Ironically, the federal government exempts tobacco manufacturers from the labeling requirements that other food and drug producers must observe. While tobacco smoking was widely enjoyed before 1880 change in the curing process make the tobacco smoking more popular after this year. Before 1880, most tobacco was cured by direclty placing it over the hot smoke of a charcoal fire. After 1880, farmers began to use hot air brought to the drying rooms by flues. This indirect drying produced a milder form of tobacco than that produced by smoke drying.

Mild tobaccos were perfect for cigarettes, a relatively "general" way of smoking. However, thee light tobaccos produce an acidsmoke when they are burned. Tobacco cured directly tends to produce an alkaline smoke. Acid tobacco smoke has little physiological effect on the body if it is simply puffed. Consequently, it must be inhaled in orde rto produce an exhilarating effect. The surface of the lungs neutrilized the inhaled smoke and the lung membrane absorbed the nicotine carried in the smoke. Smokers who inhale tend to become addicted to tobacco because they become physiologically dependent on the strong sensory reaction produced when nicotine is absorbed. Pipe and cigar smokers can become accustomed to smoking and dearly miss their habit if they quit smoking, but, in general, there is little physical reaction to the absence of nicotine in their blood since they normally do not inhale.

Nicotine acts as stimulant of the central nervous system as it is a major alkaloid in tobacco. It causes nausea, dizziness, and hallucinations in large doses. It is physiologically addicting, and withdrawal symptoms appear when it is kept from habituates. In pure form, nicotine is a poison and is often used, after treatment with sulfuric acid, as an insecticide. In addition to nicotine, smoking leads to the inhalation of tars. These tars are known carcinogens, and cigarette comapnies have therefore tried to manufacture cigarettes that reduce the intake of tars. However, none of the various filtering or extraction devices is perfect, and some tars are drawn in by smokers who inhale. There is a direct correlations between smoking and lung cancer and heart disease all these are dut to toxicity of nicotine and tars.

Smokers age prematurely, die at a younger age than nonsmokers, and tend to be susceptible to emphysema, bronchitis, and other respiratory ailments. Smoking during pregnancy increases the risk of miscarriage and problesm in early infancy. Infants born to mothers who smoke can emerge into the world underweight and addicted to nicotine. Despite the warning of the Surgeon General that smoking is bad for your health, tobacco smoking is a legal form of drugs ingestion. In direct contrast to its stance concerning many less harmful drugs, the government continues to subsidize nicotine drug addiction by providing price supports for tobacco farmers. Cigarette companies also blissfully ignore the effects of tobacco smoking and continue to lure people into smoking with advertisements that portray smokers as healthy, particularly masculine, independent, and/or beautiful.

Plant Poisons

A large number of plants are poisonous to varying degree that an enumeration would be futile. Regional agricultural stations usually publish lists of local plants that are dangerous to humans or livestock, and many horticulture books indicate which ornamental plants can be toxic if eaten. Some of these plants have been used by humans for their poisonous compounds. These are the ones in which we are interested here. As we explain earlier, a large number of plants used for their physiological effects are poisonous if used in large quantities.

It is difficult to know exactly how humans discovered which plants were or were not toxic and the quantities that could be ingested relatively safely. Ancient Greek and Roman medical rocords attest to an early interes in documenting the occurrence and usefulness of plant toxins. Administration of plant poisons as a form of capital punishment is well known from the story of the death of Socrates, after being sentenced to drink the juice of poison "hemlock" (*Conium maculatum*, Apiaceae). In the Middle Ages, "succession powders" were deviously employed to eliminate potential heirs to the throne. The employment of these powders became so common that th wealthy and powerful hired tasters as protection against sudden death from "food poisoning."

Today, scientists often seek out plants known by natives to be poisonous. Many poisonous plants are the sources of drugs or commercial poisons for use as insecticides, herbicides, or fungicides because all these poisons are either alkaloids or steroids. Some modern primitive tribes in the tropics use arrow or fish poisons such as curare or barbasco. Curare is made from extracts of the bark and stems of *Chondodendron tomentosum* (Menispermaceae) and from seeds of species of *Strychnos*, particularly *S. nux-vomica* (Loganiaceae). *Chondodendron*, a vine native to South America, contains tubocurarine. Species of Strychnos occur natively in South America, southeast Asia, and Indonesia and many yield the potent alkaloids curarine, strychnine, and toucine.

Curare alkaloids, primarily tubocurarine chloride, have had limited applications in medicine because they block neuromuscular activity and thus help relax muscles that remain contracted under anesthesia. Initially Strychnine was used medicinally as a central nervous system stimulant but has now been replaced by safer compounds. Barbasco and tuba are fish poisons used by Amazonian peoples. The first is prepared from various species of *Derris*, such as *Derris elliptica*, and the second from *Lonchocarpus nicou* (both Fabaceae).

Recently, species of both these genera have been shown to be effective insecticides because they contain rotenone. This isoflavanoid compound is fatal to insects but comparatively harmless to humans because it leaves no residue. Pyrethrum (*Chrysanthemum coccineum*, *C. cinerariifolium*, and *C. marschallii*) and tobacco acts as the sources of natural insecticides. Insecticidal compounds form species of *Chrysanthemum* are sold as Persian insect powder or Dalmation. Pyrethrins, the active compounds, are esters that break down into acids and alcohols. Since they stun insects, but do not instantly kill them, pyrethrins are usually mixed with other substances that ensure death of the insect. In some regions, notably Kenya, Tanzania, and parts of northeastern South America, pyrethrum is grown commercially for natural compound extraction. Approximately 1,200 plants are use as sources of insecticides. Yet, the fact that these would be natural insecticides does not mean that they are, or would be, less harmful than synthetic compounds. In fact, nicotine sulfate, a commercially available, natural insecticide is one of the most toxic pesticides on the market today.

13

Gums and Resins

Gums

Gums are the product of gummosis process in which breakdown of cellulose takes place. They are rich in sugars and are very much similar to pectins. They are soluble in water. There are a number of gums out of which gum arabic, gum tragacanth, and karaya gum are very important.

Karaya Gum

It is also known as India gum since it is obtained from a central Indian plant, *Sterculia ureus* of the *Sterculacea*. Other manes of karaya gum are kadaya, katira, kuteera, katilo, kullo, Sterculia gun and Indian tragacanth, The plant is a large tree. Large incisions are made into the heartwood, from which the gum oozes. Being a cheap gum it is widely used in the textile, cosmetic, cigar, taste and food industries.

Ghatti Gum

It is derived from the stem of *Anogeissns latifolia* of the family *Combretceae*. The plant is native of India and Sri Lanka. The plant is a large. The gum is used as a substitute of gum arabic.

Guar

It is obtained from the guar plants, *Cyanapsis tetragonoloba* of the *Leguminosae*. It grows in India.

Locust Bean Gum

It is variously called carob gum, St. John's bread, gum hevo, gum gatto, lupogum, rubigum, tragon, tragosol, etc. The gum is the exudates of the fruit of *Ceratonia siligna* of the *Leguminosae*.

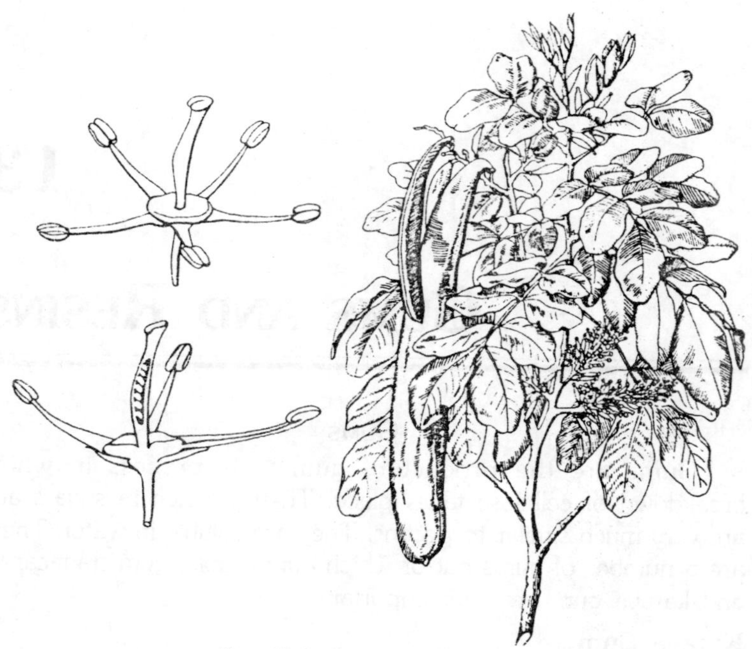

Fig. 13.1. *Ceratonia siliqua.*

The carog tree is widely grown in the Mediterranean region. The gum is a hemicellulose. The gum is used in paper making, and as stabilizer in food products.

Gum Arabic

It is obtained from wild species of *Acacia* particularly from *A. senegla* (*A. arabica*). The genus in native of Northern Africa. The plants grow wild in the dry, sandy, thorn scrub regions of North Africa, Arabia, and India. They are also cultivated in Sudaa. The plant is a small thorny tree. The trees are trapped after the rainy season by making incisions in the bark. The hardened exudates are collected at intervals of 10 days. Transluscent, tasteless and odourless gum dissolves in water completely yielding a substance that is highly viscous. The substance has a high degree of adhesiveness. It is also used in the preparation of polishes, inks, sizes, confections, medicines, etc.

Gum Tragacanth

It exudes naturally or after incisions are made into the bark of *Astrgalus gummifer* and some other species. It belongs to the

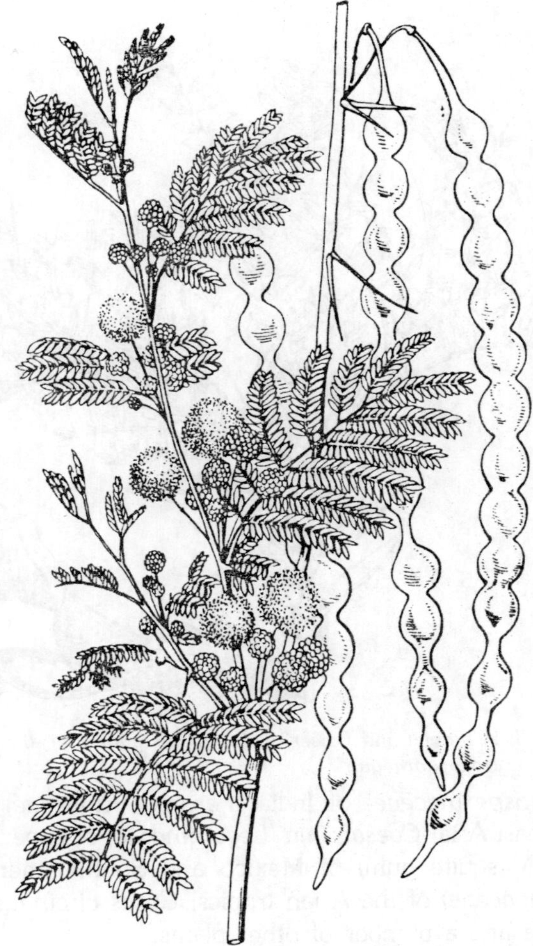

Fig. 13.2. A branch of Acacia senegal, the small tree that provides us with gum arabic.

family *Leguminosae*. The plant is native of western Asia and southeastern Europe. Iran and Turkey are the chief producers of this gum. Cell of the pith and medullary rays of the lower trunk or roots produces the gum tragacanth which is typically opaque is appearance. The gum is used in calico printing, as an adhesive, and in medicines.

Other Gums

Some of the other sources of gums are *Chukrassia* (*Meliaceae*), *Bulea* spp. (*Leguminosae*) and *Cochlospermm*

Fig. 13.3. A branch and flower of Astragalus gummifer, the major source of tragacanth gum.

(*Cochlospermaceae*) of India; *Feronia* (*Rulaceae*) of India and Southeast Asia; *Caesalpinia* (*Leguminosae*) of Argentina; *Prosapis* spp. (Mesquite gum) of Mexico and South America; *Moringa* (*Moringaceae*) of the Asian tropics; *Cycas circinalis* (Cycas gum) of Asia and a number of other plants.

Resins

Resins have been used by the ancient civilizations in the remote past. Mummy cases are coated with resins by the Egyptians. The Greeks and the Trojans used turpentines. The early Chinese and Japanese artistes used lacquer in the paintings. The manner of resin formation is not clearly known. It is belived to be oxidation products of essential oils. It is supposed that the reduction and polymerization of carbohydrates (starch, etc.) produced some resins. They may be a secondary product of plant metabolism.

Resins occur as secretions in a special ducts and canals indifferent parts of the plants. Sometimes they ooze out through the bark in a natural manner. Commercially, they are tapped by making injuries in the stems or roots of the plants. They are also obtained from fossil materials. They are insoluble in water but are soluble in ether, alcohol and other solvents. They are almost transparent. They are combustible. Chemically they are very different from the carbohydrate of the gums. They consist of resin acid, anhydride and ester mixtures, They may exude in the pure state or in combination with essential oils, gums, etc.

The families *Pinacese*, *Leguminosae* and *Dipterocarpaceae* are the sources of the commercially important resins. The utility of resins to plant is not clear but they are believed to reduce the rate of evaporation and to prevent the decay of the plants because of their antiseptic properties. Resins are, however, of great importance in industry because of their several properties. As they do not dissolve in water and harden gradually, they are used as varnishes. They are mixed with oils or solvents to impart a brilliant, transparent, waterproof layer on the painted surface. Spirit varnishes preferably with hard resins are used as paper size, furniture and leather finishes, sanding sealers and shellac substitures. The resins dissolve in alkalies to form soap.

Resins are also used in making oil cloth, linoleum, floor coatings, medicines, fireworks, plastics, incense, perfumes and in many other items. Nitrocellulose is a derivative of cellulose. It is a synthetic resins, is now widely used as spirit varnishes. The synthetic resins synthesized from phenol and formaldehyde, glycerin, urea, vinyl materials, and esters of various other organic compounds are used in paints, varnishes, safety glass, waterproofing of cloth, etc.

The natural resins are now finding less use in face of competition from synthetic resins. It is difficult to have a proper classification of resins because of the use of a variety of terms in the trades, viz., varnish resins, hard resins, spirit varnishes, balsams, gum resins, damar resins, soft resins, gums, etc. The resins can, however be classified in the following three large group :

Hard resins — Usually solid
Oleoresins — Usually liquid
Gum resins — Mixture of gums and resins.

Hard Resins

The resins containing very less essential oils are considered as best sources of varnishes and hard resins are one of them.

Moreover. They dissolve in alcohol. They are usually solid, tasteless and odourless. They are combustible and burn with a smoky flame. Copals and damars, the two most important commercial resins, fall under this category. Hard resins besides being very good sources of varnishes are also used in paints, plastics, inks, fireworks, sizing adhesives, etc.

Amber

It is one of the oldest resins in use. It is hard and brittle. It is obtained from the fossils of *Pinus succinifera*, found along the shores of the Baltic Sea. Amber is yellow, brown or black coloured. Dolls and ornamented items are formed by ambar. It is also used for the mouthpieces of pipes and holders for cigars and cigars and cigrettes. It also yields varnish and an essential oil. It is used to increase the elasticity of rayon fibres.

Shellac

It is not produced by plants, but from an insect *Laccifer* (*Tachardia*) *lacca* which feeds on *Butea monosperma*, *Schleichera oleosa*, *Zizyphus xylopyrus*, *Ficus religiosa*, *Acacia nilotica*, *Cajanus cajan* and *Zizyphus jujuba*. These plants are culitvated in India on a large scale for obtaining shellace (97%) of the total world output). Burma, Thailand and Indo-Chinese countries account for the rest.

Shellace is widely used in industries. It's characteristic to be easily moulded into various shapes. It is very useful in the manufacture of phonograph records, insulating materials for electrical goods etc. spirit varnish is obtained from it. Spirit varnish is used for interior works science it is not water-resistant. It is also used in the manufacture of sealling wax, drawing inks, water colours; for stiffening felt hats and for sizing papers and for decorative purposes. Other hard resins are acaroid resins (*Xanthorrhoea* spp.), sandarac (*Tetraclinis articulata*), mastic (*Pistacia lentiscus*) Dragon's blood (*Daemonorops* spp), kinos (*Pterocarpus* spp.)

Copals

They are groups of very hard resins. They are obtained from fossils and semifossils.

Kauri Copal. It is obtained from the Kauri pine (*Agathis australis*) of New Zealand. The plant belongs to *Pinaceae*.

Manila Copal. It is derived from *Agathis alba* of *Pinaceae*.

West African Copals. They comprise *Cango copal*, *Sierra copal* and *Acera* and *Benin copals*. The Congo copal and Sierra copal are obtained from several species of *Copaifera* of the *Leguminosae*. Acera and Bonin copals are derived from *Daniella ogea*.

East African Copal. They comprise *Zanzibar copal*, *Madagascar copal* and *Mozambique copal*. They are all derived from *Trachylobium verrucosum* of the *Leguminosae*.

South American Copals. The are derived Hymenaea courbaril of the *Leguminosae*.

Lacquer

It is obtaines from the lacquer tree (*Rhus verniciflua*) of China and Japan It is a very good varnish which is used for ornamental purposes. *Burmese lacquer* is obtained from *Melanorrhoea usitata*.

Damars

They are obtained from members of *Dipterocarpaceae* growing chiefly in malaysia and Indonesia. The damars are as follows :

Temak Damar is obtained from *Shorea hypochra*.

Batavian Damar is obtained from *Shorea weisneri* of Java and Sumatra.

Penak Damar is obtained from *Balanocapus heimii*.

Mata Kuching Damar is obtained from *Hopa micrantha*.

The Indian domars are sal damar (*Shorea robusta*), white damar (*Vateria indica*) and black damar (*Canarium strictum*).

Oleoresins

Oleoresins are liquid in nature due to the presence of essential oils. They have characteristic odour and flavour. Some of the important oleoresins are turpentines. balsams and elemis.

Turpentines

Coniferous trees are the source of turpentine. The crude turpentine or pitch obtained by tapping the trees is distilled to yield an essential oil, oil of turpentine or *spirits of turpentine*, and *rosine*. The major resin canals are interconnected and run vertically in the sapwood, while smaller ones connect them to the bark. The turpentine industry is also called 'naval stores' industry because of the widescale use of turpentine in the sailing vessels from early days. Turpentine-yielding plants are *Pinus palustris* or *australis* (90% of total yield), *P. caribaea*, *P. pinaster* and some other species of *Pinus*.

Turpentine gives spirits known as turpentine which is widely used as a solvent and thinner in paints and varnishes, in medicine and as a solvent in rubber and guttapercha. It is also used in the printing cloth and in the manufacture of many chemicals.

Rosin or colophoney is a solid. It has many important industrial uses. It is used in the manufacture of soap, paints, varnishes, waxes, seals, lubricants, inks, coatings, oilcloth, linoleum, plastics, rubber, drugs, chemicals and adhesives, etc.

India, Indonesia and Indo-Chinese countries are the chif producers of turpentine and rosin. An Indian non-coniferous tree, the Indian frankincense (*Boswellia serrata*) also yields turpentine.

Other turpentines of lesser importance are Canada balsam (*Abies balsamea*), spruce gum (*Picea rubens*), venetian turpentine (*Larix decidua*). *Bordeaux* turpentine (*Pinus Pinaster*) and some others.

Balsams

They contain less oil than turpentines and are, therefore, thick, syrupy viscous liquids. They are highly aromatic due to the presence of benzoic of cinnamic acid. It has medicinal uses for dressing and healing of wounds and in coughs, colds bronchitis, etc. in the perfume industry as fixatives, as adhesives and flavourings, etc. Some of the examples of balsams are balsam of Peru (*Myroxylon balsamum*), balsam of Tolu (*M. bolsomum* var, *periiae*), styrax (*Levant styrax*), benzoin (*Styrax* spp.), copaiba (*Copaifera* spp.), etc.

Elemi

They are liquids at the time of exudation but harden on exposure. They comprise a number of different oleoresins. Manila elemi is obtained from the pili tree (*Canarium luzonicum*). African elemi is obrained from the *Boswellia freriana*. Elemis are used for torches, caulking boats; in the manufacture of adhesives, inks, cements, in varnishes, medicines and perfumes.

Gum Resins

It is explained earlier about gum resins. Gum resins are the mixture of gums and resins. The exude as milky substance and harden into small irregular masses.

Ammoniacum is obtained from *Dorema ammoniaeum* of the *Umbelliferae*. It is used in perfumes and as a stimulant of the circulatory system.

Asafetida is obtained from *Ferula astafoetida* of the *Umbelliferae*. It is greyish or reddish in colour. It has a foul smell and a bitter acid taste due to the presence of sulphur. It is used for flavouring curries, sauces and other edible items. It is also used in medicines for cold, coughs, asthma, disgestive disorders and in perfumes.

Galbanum. *Ferula galbaniflua* of the *Umbelliferae* is the source of Galbanium. It has strong smell and is used in medicine.

Myrrh. This gum resin is obtained from *Commiphora* spp. of the family *Burseraceae*. It is used in perfumes, incense, ointments, medicines, disinfectant and embalming mixtures.

Frankincense is obtained from *Boswellia* spp. of the *Burseraceae*. It is used as ceremonial incense in the Church, and in perfumes, face powders, pastilles, etc.

14

SMOKING MATERIALS

Man has been using smoking and chewing materials since times immemorial for the pursuit of pleasure. Presence of alkaloids give most of these materials a stimulating or narcotic effect. The narcotic plant pruducts and their derivatives are extensively used the world over. Their continued use make people drug addicts. These materials have very harmful effects on them. Senses can be lost temporarily and sometime permanently as a result of addiction. Such people are problems for family as well as for the society.

Tobacco (Nicotiana tabacum)

Tobacco is extensively used throughout the world for smoking and chewing purposes. Dried, cured and fermented leaves are used because of the stimulating and slightly narcotic effects.

Origin and history

Tobacco is indigenous to tropical America. The plant has been under cultivation from ancient times. Its original wild ancestor is not known. the plant was introduced into North America long before Columbus discovered it. He saw the natives of Cuba using tobacco in 1942. It was introduced into Spain in by Spanish explorers in 1558. It was carried to England in 1585 by *Sir Walter Raleigh*. By the end of the sixteenth century and early seventeeth century the plant spread of other part of Europe, Africa, India, China, Sri Lanka, and Australia. Tobacco is now cultivated in U.S.A., India, China, Rhodesia and parts of East Africa, the Balkan States, Russian, Japan and Brazil.

Fig. 14.1. Species of tobacco, *Nicotiana tabacum.*

Botanical description

Tobacco plant belons to the family *Solanaceae*. It is a small herbaceous annual. In the early stages it appears is a rosette but grows to a height of three to five feet. Top root system of the plant is quite strong and well developed. The stem bears very large ovate leaves. Flowers are white or pink borne in receme inflorescence. The capsules contain numerous small seeds.

Commercially two species of Nicotiana are important—*N. tabacum* and *N. rustica* of which the former is by far the more imporant one. *N. rustica* occurs wild in Mexico and in cultiated to some extent in part of Europe, Northern Asia and Indonesia. Having high content of nicotine, it is mainly grown as a source of drug and for insecticidal purposes.

Fig. 14.2. Species of tobacco, N. rusticum.

Ecological factors

Tobacco is a tropical crop but grows well in sub-tropics and even in temperate climates. Moderate rainfall evenly distributed throughout the year is good for the crop. The crop thrives best in a well-drained light sandy loam. The soil with rich humus and potash, lime and other essential elements is good for the plants.

Cultivation

Broadcast method is employed to sow tobacco seeds. The seedlings are transplanted in well cultivated fields. Topping, process of removing tip of the plant when it produces eight to twelve leaves results in development of axillary shoots. These shoots are then removed to encourage increase in the size, weight, thickness and strength of the leaves. It takes four to five months for leaves to mature. The leaves turn greenish yellow during ripening.

The leaves are harvested as they ripen from below upwards for giving best quality tobacco. Very often the whole plant is harvested. Different methods of curing determines the time of harvesting with different degree of ripeness of the leaves. The leaves of tobacco contain about 70 to 80 per cent water, starch, nitrogenous compounds, organic acids and 1 to 8 per cent of an alkaloid, nicotine.

Curing of tobacco

The smell and taste of tobacco develp during the curing process due to the liberation of varous essential oils and other aromatic substances. The harvested plants or leaves are stacked or hung in curing barns. Curing is of four kinds. *Air-curing* is done in well ventilated barns under moist and warm conditions. It is a slow fermentation process. *Flue-curing* is a faster process. It involves heating of the barns with the help of furnaces. Smoking from slow fires of charcoal etc. with no significant temperature rise is the method for *fire-curing*. *Sun-curing* is performed in the open in the Asian coutuntries.

Curing involves dry fermentation of the leaves which lose water, green colour, starch and sugars. Protein (about 60%) is lost in the form of ammonia gas. The amount of oxalic acid is not changed but citric acid increases. The volatile form of nicotine evaporates and, therefore, the amount of nicotine (only the non-volatile form remains) decreases. Curing takes three to six months'

Fig. 14.3. Indians smoking a tabac (pipe).

time. The cured leaves are then sorted and similar grade leaves are tied into 'hands'. The graded leaves are placed in heaps or pressed into special containers for the purpose of fermentation. The fermentation is done in large warehouses. It takes six months to four years' time to complete the process during which chemical composition of leaves is changed further.

The nicotine content is further decreased. The characteristic odour and flavour of the tobacco develop during this period. The harshness, bitterness and othe rundesirable qualities disappear and the burning qualities of the tobacco is enhanced. Chewing, snuff and smoking are the various ways in which tobacco is used. Tobacco is smoked in a nubmer of ways.While the villagers use earthen pots called '*Chilam*' and '*Biri*' for smoking tobacco, the urban people use cigaretters, cigars, pipes etc.

Fig. 14.4. Mayan priests blew smoke to the four winds during religious ceremonies.

Betal and Areca Nut (Piper betle and Areca catechu)

Betel or *pan* is a very popular chewing material in the Asian countries. It is used extensively by the people of Zanzibar, India, Burma, Malaysia, Indonesis, Vietnam, Thailand, Cambodia, South-China and the Philipines from the remote past. An average Indian particularly in the North is very fond of betel and consumes it several times in a day. Betel has four pincipal ingredients—leaf of betel paper (*Piper bettle*), seeds of Areca palm (*Areca catechu*), lime and a resin called 'katha'.

The betel pepper is a perennial, creeping out climbing plant. It is a native of Malaya and is cultivated throughut the south-eastern Asia. The leaves are dark green and having a hot astringent taste. They are broad with pointed tips.They are borne alternately on the stem. The plant is grown under shaded conditions to cause an increase in the size of the leaves. The Areca palm is a very tall slender palm. It is grown throughout Asia for its seeds. There is a crown of large pinnate leaves at the top of the tree. The inflorescences consist of much-branched spadix. The fruit is an ovoid drupe. The ripe fruit is orange coloured. The pericarp is a hard fibrous structure. It encloses a single seed.

An alkaloid arecoline is present in seeds. Fresh or cured seeds are used after slicing them into small pieces. Cured seeds are better than fresh seeds because they have lower tannin content, better palatability and better storage properties. Pan consists of a leaf of betel vine with a mixture of areca nuts 'katha', and staked lime rolled in it. Betel is often chewed along with some other aromatic substances as well. These are cloves, fennel cardamom, etc. Tobacco is also quite often used as an ingredient of betel. Betel is chewed slowly in the mouth. It causes flow of saliva which is spit out every few minutes. It has stimulating and mildly narcotic effets on the human body. It is also believed to have some medicinal value.

Cola

The cola or kola nuts are the seeds of the cola tree (*Cola nitida*) of the *Sterculiaceae*. They are widely used as a chewing material as well as a source of flavour and caffein in 'cola' drinks. The cola tree is indigenous to tropical western Africa but is now grown in Sudan, Jamaica, Brazil, India, etc. It is a forest tree. Shar-shaped fruits contain 8 reddish seeds. Fresh cola nuts are known to have stimulating effect when chewed, and inhibits fatigue and hunger.It contains per cent caffein, an essential oil, and a glucoside–kolanin. Stimulating effect of cola seeds is due to the presence of caffein and kolanin.

The True Narcotics

There are some plants which contain alkaloids of great economic importance. In very small amounts these alkaloids are used in medicines for relieving pain, anxiety, tension, etc. and to induce sleep. Cocain and opium ar examples of such narcotics. Hashish, peyote cappi, etc, are however, very strong narcotics. They have hypnotic and halucinating effect on the mind of the user. They excite the brain to so much extent that the person sees unreal visions and illusions. The drug habit results in great physical and mental degradation, senselessness, coma, convulsions and absurdity.

Coca

The leaves of coca plant (*Erythroxylon coca*) are used for obtaining the drug *cocain* and also for chewing purposes. The plant is a shrub. It is native to peru and Bolivia. It is now grown in South America, Java, India, Sri Lanka and Formosa. The leaves mixed with lime and alkaline ashes of some plants are chewed to stimulate the physical and mental activities. The hunger and pain etc. are forgotten under its influence. Loss of health disease and disability results due to its excessive chewing.

Opium

Opium has been dealt with as medicinal plant in the previous chapter. Opium smoking and eating has been prevalent in the Asian countries for centuries. It is now used in large number of countries. It is used for its strong narcotic effects. It is very dangerous drug containing of alkaloids morphine, codeine, narcotine and papaverine. Opium is the dried juice which exudes from the immature capsules of the opium popy (*Paraver somniferum*) of the *Papaveraceae*. The plant is native of Asia Minor of India. It is a herbaceous annual and flowers in three months' time only. While Indians eat opium, the Chinese smoke it. Its worse effect on human body is exhibited when it is inhaled as fumes. The alkaloids, morphine and codein are extensively used in medicines for the relief of pain. The seeds of poppy are edible. They contain 4 to 50 per cent of a drying oil which is used in India for the preparation of sweetmeats and in curry. The oil is also used in the manufacture of paints and soap.

Ganja, Bhang (Mrijuana, Charas, Hashish)

The hemp plant (*Cannabis sativa*) has three kinds of varieties under cultivation–for ribres, for a drying oil and for a narcotic drug. The durg yielding variety is extensively cultivated in India, Nepal an the West Asian countries for thousands of years. The Chinese are reported to have used the drug in 3000 B.C. The resinous hairs of the inflorescence of female plant are the source of the narcotic stimulant or the drug or the resinous oil. It is used for relieving pain and for treating nervous disorders. The drug contains powerful alkaloids.

The drug is prepared in different ways in different part of the world. It is smoked as well as consumed as a beverage. *Ganja* is produced from the dried and powdered young female inflorescences and young leaves. It is usually smoked. *Bhang* or *Marijuana* (American name) is consumed along with water, milk or sweetmeats. It consists of tops of wild plants. Female inflorescence before the fertilization takes place exudes a resinous secretion which gives highest concentration of the drug. Charas and hashis (West Asian name) are the pure resin. The active principle is tetrahydrocannabinol.

The various forms of the hemp listed above all are extremely dangerous narcotics for the human beings. They produce very deleterious psychological effects. It causes great amount of cerebral excitement, hypnotic effect, haliucinations, erotic visions, imaginary flights into the world of unreality, mental unconsciousness, stupefaction, ecstasy and the like. The modern society of the western countries is today plagued with such addicts.

15

GRAINS

Life does not exist only on bread so approximately 70% of the world's farmlands are used to the growing of grains. Harversts from these cereals provide at least half of the world's calories today. Not only humans, but there domesticated animals as well are fed primarily by grains and forage grasses. In this chapter we examine these remarkable plants and discuss why grasses have become the most important group of food plants in the world today. First of all grains re cultivated among all the cultivated crops. In the opinion of some anthropologists, their use was a prerequisite for civilization.

Wheats and barley sustained early Near Eastern cultures, rich the Far Eastern populations and corn the pre-columbian new world civilizations. Contrary to the popular view that Rome fell because of decadence and debauchery, its decline can be attributed, in part, to the failure of its North African grain supplies. Failures in grain cultivation have destroyed so many other cultures in the past and threaten modern economies as well. By definiation, grains are the fruits of members of the grass family, Poaceae. They are indehiscen,t one-seeded fruits in which the seed coat is fused with the ovary wall. Grain amaranths and buckwheat, often grouped with the grains, are fruits of species belonging to the Amaranthaceae and Polygonaceae, respectively. Consequently, they are not true grains.

Grain yielding grass are known as cereals, in deference to the Greek goddess Ceres. The grass family contains over 9,000 species distributed worldwide, but only an estimated 35 have been cultivated as cereals. Of these, less than a dozen are important today. These were 35 species at some time. Out of these 35 species, only five were domesticated in the New World, and of the top half-dozen major cereals in the world today, conrn is the only New World native. While the natural characteristics of some wild grasses encouraged their initial domestication by humans, thousands of years of artificial selection have exagggerated the differences between cultivated and wild forms. Yet, within grains as a group, we can trace several directions of human selection depending, in part, on the characteristic of the original wild species. In order to appreciate why people aroudn the world chose grasses as their primary food crops and how, under the action of human selection, grains rose to their modern preeminence as mainstays of the world's populations, it is necessary to understand the biology of grasses.

The Grass Plant

Grasses are ubiquitous. In nature, the matted roots of wild grasses hold down soils from the Arctic tundra to tropical savannas, while planted grasses carpet our manicured suburban lawns. Yet, perhaps because of their green, inconspicuous nature, we never seem to notice grasses, even though they are constantly underfoot. When these are examined very closely, this group of monocotyledons proves to be elegant in design and different in many respects from the "generalized" plant. Grasses have fibrous roots but lack of taproots. Grasses also have a different pattern of growth from most other kinds of plants.

Many species branch only at the base, producing more or less equally sized vertical stems called tillers. Still other grasses have one or a few major stems that produce sequential branches as they grow. Individual stems or tillers are made of fused, clasping leaf bases surrounding a thin growing axis. Although we tend to call only the flattened, divergent part of a grass blade "a leaf" the leaf actually extends for a considerable distance below this point. The regions along the stem where leaves differentiate from it are called nodes. Somewhere above the nodes, the leaves bend outward. At this point there is often a rim of hairs or a membranous structure called the *ligule*. The reproduction is vegetative in most of the perennial grasses.

Many species have rhizomes that grow outward as the grass clump ages. Still others produce horizontal branhces (*stolons*) that

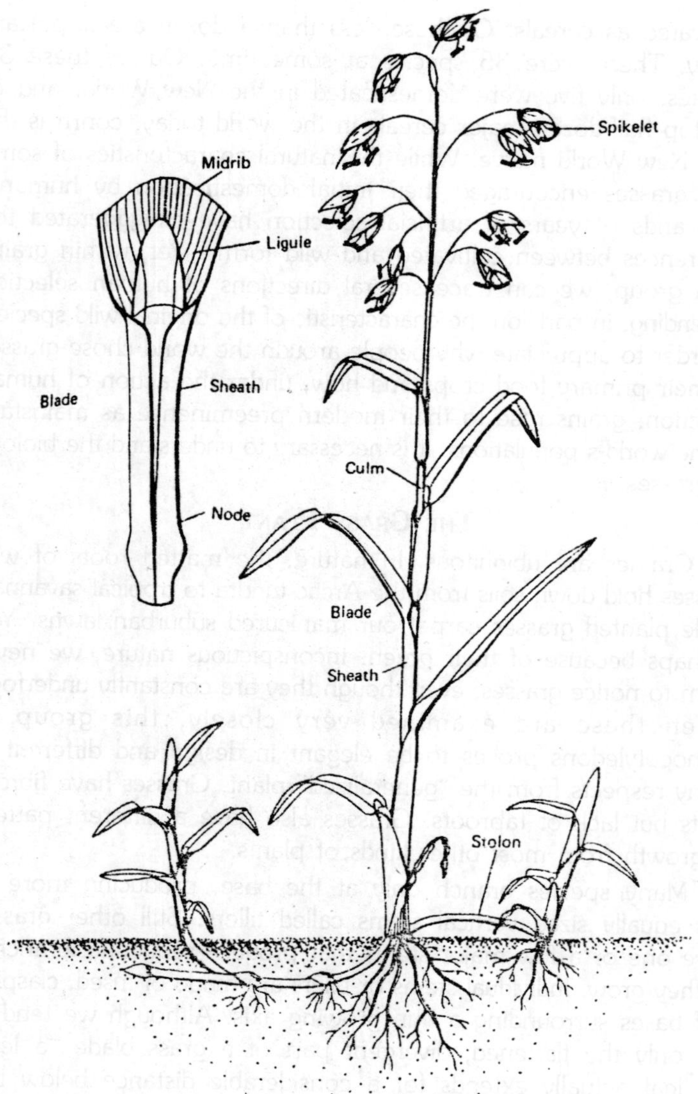

Fig. 15.1. A generalized grass plant.

grow along the ground, often rooting at the nodes as they go. A stoloniferous vegetative growth permits the rapid spread of grasses that are inserted into lawns in the form of starter plugs. Grass floral structures are even more highly specialized than their stems and leaves. Most grass flowers are perfect, although those of corn

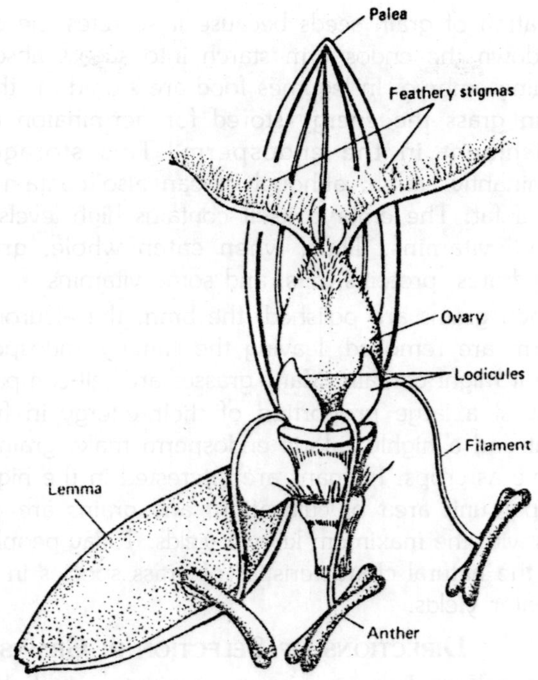

Fig. 15.2. A generalized structure of a grass flower.

and wild rice are notable exceptions. Grass flowers are borne in compound inflorescences composed of series of *spikelets*. Each spikelet is flowering branch with one to several florets.

A *floret* is an individual grass flower which typically has a superior ovary topped by two feathery stigmas, three stamens (sometimes six), and three scalelike remnants of the petals. Each floret is enclosed in two protective bracts. The inner bract is the *palea*, the outer is the *lemma*. The nerves of the lemma can extend beyond the body of the lemma, forming a long structure called an *awn* that aids in dispersal. At the base of each spikelet (or cluster of florets) are additional (sterile) bracts called *glumes*. After fertilization, the single ovary matures and develops into a fruit in which the seed coat and the ovary wall (pericarp) are fused into a complex structure known as the *bran*.

The different components of the bran can be distinguished only microscopically. Inside the bran is a layer of cells known as the *aleurone layer*. The cells of this layer are typically rich in protein and fats. The aleurone layer plays a key role in the germination of grain seeds because it secretes the enzymes that break down the endosperm starch into sugars absorbed by the expanding embryo. In legumes food are stored in the cotyledons while in grass the energy stored for germinaion and seedling establishment in the endosperm. This storage produt is predominantly starch, although it can also contain protein and traces of fat. The embryo itself contains high levels of proteins, fats, and vitamins. Thus, when eaten whole, grains contain carbohydrates, proteins, fats, and some vitamins.

When grains are polished, the bran, the aleurone layer, and the germ are removed, leaving the starchy endosperm and any protein it might contain. Many grasses are self-compatible annuals that invest a large proportion of their energy in fruits. A high fruit set and a highly caloric endosperm make grains particulalry attractive as crops. Humans are interested in the highest possible yields per unit area of cultivation, and grains are among those species with the maximum known yields. Today people have made use of the natural characteristics of grass species in the selection for greater yields.

Directions of Selection in Grains

There have been two major ways in which humans have modified thehabit of cereals to increase the total amount of grain that can be easily collected at one time. First, in grains that ancestraly had a tillering habit, we have selected for synchrony of tiller formation. Since each tiller is terminated by an inflorescence, tillers that begin growth at the same time tend to mature seeds. Thus, by replacing sporadic tiller formation with simultaneous tiller initiation, we have been able to ensure that a single harvesting of grains such as wheat, rye, and barley would collect almost all of the seeds produced. Grains having branched stem, the lateral branches as well as main stem usually bear one or more inflorescences. Since the plant grows upward, lateral branches are produced sequentially from the base to the tip. Thus it is normally impossible for all of the lateral shoots and the main shoot to mature fruits at the same time.

In cereal crops that ancestrally had branched primary stems, there has been a selective trend for the elimination of branching and the production of single-stalked individuals with high fruit production per stalk. Modern cultivars of corn, sorghum, and pearl millet all have a large, single, unbranched stem, wheereas their ancestors or early cultivars were branched. There has also been selection in recent times for short plant stature. When the grains are grown in areas much wetter than their native habitats it causes *lodging*. Lodging is the matting together of stalks which have been bent over by wind or rain. The matted stalks can not disentangle and return to an upright position. Once lodging has occurred, the stalks geneally rot.

The native areas of the Old World cereals are relatively dry, at least during the growing season. When humans tried to grow these grains in wet areas of the world, lodging became a serious problem. Intensive selection programs for increased stalk strength and for dwarf varieites of wheat, rice and even sorghum, have allowed a dramatic increase inthe cultiavation of these cereals, primarily in the tropics. Under the influences of human selection, the inflorescences of the cultivated cereals have undergone even more drastic changes than the vegetative portions. One of the initial changes in the cultivated cereals was the retention of fruits on the inflorescence once theymatured. The dispersion of seeds to areas suitable for germination because the main problem for all flowering plants. Many grasses disperse their seeds by having inflorescences that become brittle at maturity so that they shatter when jostled by wind, pressure from passing animals, or neighbouring plants.

From the human point of view, shattering is an undesirable character because the seeds fly in all directiosn when harvesting is attempted. Any mutation, or series of mutations, that dampended or prevented shattering would have been selected because seeds of the nonshattering type would have preferentially been collected and subsequently used as seed sources for the next year's planting. Nonshattering is a genetically controlled character due to which over time it would become the predominant type used in cultivation. Whether or not the trait was intentionally selected for by humans, all of the major cereal crops now have nonshattering inflorescences.

Another major change in the morphology of grains during their domestication is related to the ease of fruit separation from the inflorescences once the stalks have been cut. Grains can separate from the stalks below the attachment of the bracts, or the fruits can fall free from the enclosing bracts (chaff). Primitive and wild grains come free from the stalk still enclosed in their bracts. *Threshing* is a method of removing the fruits from the braets. In some grains the fruits falls free of the braets. Threshing was formerly done by hand by beating the stalks. Hand threshing is an arduous job, and any mutation that lessened the task would have been seized upon by humans. The appearance of "free-threshing" types in the archaeological record documents early selection for mutations that facilitated the separated of the fruits from the chaff. Free-threshing types solved the dilemma of needing a grain that would not shatter before harvest. but which separated easily from the enclosing bracts once collected. Winnowing is a process that involves throwing the grain-chaff mixture into air. The light fragments of broken chaff that are mixed with the threshed grain are removed by winnowing. The light-weight chaff fragments are blown away by the wind, while the heavier grain falls back down again.

In industrialized countries. modern combines thresh and winnow many grains as they are harvested in the field. but threshing and winnowing by hand are still commonly practiced in many parts of the world. The separation of any remaining pieces of the fruit wall from the seed is very difficult because of the nature of the coat of grain

In ancient times, the fruit wall was removed by rubbing the grains (wheat, rice, or occasionally barley) between two stones with enough pressure to abrade off the unwanted layers without crushing the seeds. Today, these grains are hulled, dehusked, or "pearled" to various degrees by machines. The complete removal of the joint fruit and seed coat (which generally also pulls away the embryo) is very common for wheat and rice and leads to white flour and polished (white) rice. Degermed grains can be stored longer than those with germs in this way the removal of the germ serves a purpose. The oils in teh germ tend to become rancid over time, particularly if the grain is crushed. Groats are grains, often broken. from which only a minimum amount of the fruit wall has been removed. Additional modifications of grass inflorescences include selection for increased tiller production, restoration of fertility to sterile florets. increase in individual grain size. and changes in the kinds of storage products in the endosperm. When we discuss each of the major grain crops, we point out the particularly important kinds of artificial selection to which each has been subjected.

MAJOR GRAIN CROPS

Barley

On the basis of archaeological evidence, barley (*Hordeum vulgare*) appears to have been the first cereal domesticated. Archaelogical digs in the Fertile Crescent (primarily in Syria and Iraq) haveuncovered charred barley kerenels about 10,000 years old. Wild forms of barley are all two-row and in the Fertile Crescent, the oldest cultivated forms of barley are also two-row. This type of barley has only one fertile floret in each spikelet of three on the flowering stalk. By 6000 B.C., six-row forms appear in the archaeological record. In six row types, each of the three florets in a cluster matures into a fruit. Since each node bears two

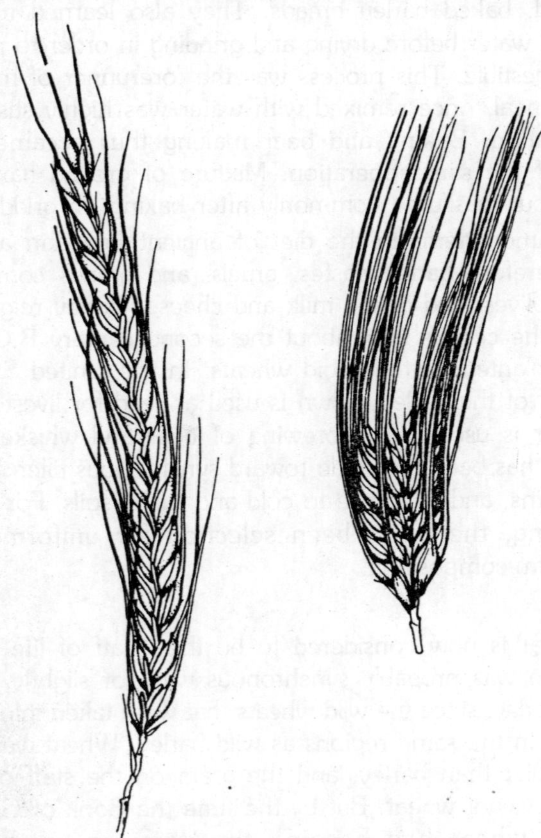

Fig. 15.3. Two- and six-row barley.

spikelets, the net effect of suppression of two of the three flowers in each spikelet is the productin of an inflorescence with two rows of grain. If all of the florets are fertile, the final infructescences bears six rows of grain. Initially, barley appears to have been ground and made into pastes. However, since raw starch, even mixed with water, is relatively unappetizing, the grain was probably first toasted by heating it on stones. The pastes might also have been boiled into porridges over coals in stone-lined pits.

By Egyptian times, barley had become such an integral part of the culture that it ahs its own hieroglyphic symbol. Likewise, the ancient Greeks depended primarily on barley and greatly expanded its uses. In additon to pastes, they produced variously flavoured, baked barley breads. They also learned to soak the grains in water before drying and grinding in order to make them more digestible. This process was the forerunner of mating. The ground meal, once remixed with water was highly susceptible to fermentation. Baking and beer making thus became part and parcel of the same operation. Mixture of ground flax seed and barley flour was used commonly after baking on griddles.

Common items in the diet of ancient Egyptian and Greeks were therefore barley pastes, gruels, and breads complemented by figs, olives, and goat's milk and cheese. Barley reigned as the king of the cereals until about the second century B.C., when it was supplanted by tetraploid wheats. In the United States more than half of the barley grown is used as feed for livestock. About a quarter is used in the brewing of beer and whiskey. Modern selection has been directede toward synchronous tiller production, short stems, and tolerance to cold and saline soils. For grain used in malting, there has been selection for uniformity of the endosperm composition.

Wheat

Wheat is now considered to be the "staff of life". Its initial cultivation was probably synchronous with, or slightly later than, that of barley, since the wild wheats that were taken into cultivation occurred in the same regions as wild barley. Wheat was originally less popular than barley, and the grain on the staff of Demeter was barley, not wheat. But by the time the Book of Genesis was written, wheat had become the dominant cereal of the Mediterranean area and was the grain from which Adam and Eve were forced to make their own bread. Wild and early domesticates of wheat were diploid (2n = 14) and have been classified as *Triticum monococcum*. Natural mutation occured at some point during the early cultivation of the species that suppressed shattering. This mutant was quickly adopted and soon become the major cultivated type.

Today these wheats, known as einkorns, are still cultivated in a few remote areas of Yugoslavia and Turkey. By the eightn centurey B.C., an einkorn wheat and another specis (though by some to be a wild goat grass (*Triticum speltoides*), but by others to be *T. searsii* or *T. longissimum*), hybridized. It is difficult to determine the species which hybridized with the einkorn wheat because the chromosomes contributed by this species appear to have undergone many changes since the initial cross. The first change which occurred was a doubling of the chromosome complement, which producted tetraploid emmer wheat (*T. turgidum*). Unlike einkorn wheat, one variety of emmer wheat (*T. turgidum* var. *dicoccum*) underwent a mutation which caused the bases of the glumes to collapse at maturity. This change made the separation of the fruit from the chaff relatively easy and thus gave rise to the free-threshing type of wheat known as durum wheat (*T. turgidium* var. *durum*).

Perhaps even more important than the free-threshing nature of these grains was the change that polyploidy made in their ability to produce raised bread. Wheat contains protein and starch. Proteins are gliadin and glutenin these along with starch combine and form a tenacious complex called as gluten. Because of these proteins, when flour from bread wheat is mixed with water and kneaded, it becomes elastic. If yeast is added to themixture, the yeast cells carry out their normal process of fermentation using the small amount of natural sugars in the flour. However, when the yeast cells release carbon dioxide as a product of fermentation, gas becomes trapped in the spongy mass and the dough rises. Baking "sets" the dough, by drying the starch and forcing the gas out of the pockets in the dough. Cooking also kills the yeast cells. Prior to the evolution of free-threshing wheats, wheat was commonly heated to facilitate threshing.

This process causes side effect because the proteins coagulated and became inclastic and unable to trap carbon dioxide. Flour made from grain treated in this way was useless for making raised bread. Wheat naturally has more gluten than other grains and is the only one capable of forming a light dough if the proteins remain in their natural state. Howver, until free-threshing types occurred. the advantages of gluten were obscured. The formation of the tetraploid emmer wheat simultaneously produced a grain with an especially large amount of gluten and the free-threshing condition. These features combined to produce a wheat that was easily separated from the bracts and capable of producing light, raised bread.

Durum wheats are known today as a hard, marconi, wheats and are primarily grown in areas of low rainfal such as the Mediterranean region, central Russia, and the northern Great Plains. These wheats are now used primarily for pasta and noodle making because another event in the history of wheat produced a grain superior even to durum wheat for the making of bread. Hexaploid species arose just after the appearance for tetraploid wheats. Hexaploid had two sets of chromosomes. one from a tetraploid emmer wheat and another set from a diploid species generally assumed to have been *Triticum tauschii*. These free-threshing hexaploids, known botanically as *T. aestivum*, have an especially hard, proteinaceous endosperm that surpasses other wheats in bread making. There are now over 20,000 cultivars of bread wheat. including the hard red and the club wheats. The common names of wheat varieties are confusing since, in some cases, the common name of one variety is the same as the cultivar name for another. In other cases, the common names refer strictly to the time of sowing of wheat. For example wheat sown in spring called spring wheat and wheat sown in fall called winter wheat.

Winter wheats germinate in the fall and survive the winter as seedlings. When spring arrives, the seedlings rapidly send up numerous tillers. Where winters are too severe for seedling survival, spring wheat is planted as early as possible for a fall harvest. The harvesting sequence of the various wheats across North America is such that combines can start harvesting wheat in May in Texas and other southern wheat-growng areas. and then follow the maturation of the crop northward until fail. Wheat has not always been a major crop in North America. The grain is brought to the New World in 1520 by the Spanish, and English colonists also tried to grow it in their early settlements. Almost all of these early agricultural attempts failed.

Mennonite immigrants in western Pennsylvania were one of the few groups to be successful with wheat in the eastern United

States. As settlers moved westward across North America, they carried seeds with them, but prior to the 1870s, wheat growing in the midwest was plagued with problems similar to those experienced in the east. Periods of drought, strong winds, and various pests took heavy tolls on the wheat crops. In the 1870s immigrants began to pour out of central Russia hoping to dodge military service or to find religious freedom.

Like many other groups of immigrants, they settled in regions similar to the ones they had left behind. In this case, they chose Kansas, North Dakota, Montana, Oklahoma, and parts of Texas. The ideal climatic conditions are available in these areas for wheat farming: temperatures averaging above freezing in the winter and an annual rainfall of 32 to 87 cm (13 to 35 in). Most of the wheat brought by the immigrants from Russia was "Turkey" wheat, although it is debated whether the variety originally came from Turkey or the Crimea. Because of its superior qualities, this particular variety made wheat farming feasible in the United States. It subsequently became the ancestor of the successful American varieties. Two later developments contributed to the success of wheat growing in the United States. One helped to alleviate pest problems, and the other led to better qualities of flour, the principal product obtained from wheat. Although selection improved many aspects of cultivated wheat, it failed until recent times to produce disease-resistant varieties. The rusts, fungi are the major wheat pests that infects the fruiting stalk and destroy the crop.

The U.S.D.A. and the Minnesota Agricultural Experiment Station began looking for ways to combat black stem rust (caused by a fungus. *Puccinia graminis*) in 1907. Chemical pesticides proved unfeasible, and destruction fo the other necessary host of the fungus, barberry (*Berberis vulgaris*, Berberidaceae), did not significantly reduce infestation. Breeders finally learned that resistance could be conferred by a change in a single gene. A large number of breeding programs begun and still continue. Agricultural agencies now collect and maintain stocks of mutants of this gene that can be used to breed resistance into current varieties. Just as resistance to rust is conferred by a single gene, the ability of the rust to overcome the resistance appears to be as simply controlled.

Consequently, mutations that permit the rust to infect previously immune wheat varieties continually arise, necessitating the incorporation of a different form of the resistance gene into the crop. By knowing the genetics of rust immunity and interjecting alleles for resistance quickly into seed stocks, the wheat become one of the major crops of the United States. The development of the modern industrial flour mill is an another important event which enhance the wheat production. Steel rollers replaced traditional mill stones and successive layers of silk bolting allowed the rapid sifting of a fine white flour.

The purpose of a flour mill is to produce finely ground wheat. To produce white flour.the starchy endosperm must be separated from the bran and the germ. In modern mills, whole grains are fed between grooved steel cylinders that move at varying speeds. The first rollers break and tear open the grains. The "chop" produced by this operation is sieved into three components: (1) first-break flour, (2) coarse nodules of starch, and (3) large pieces of the grain wall with some endosperm still attached. The coarse nodules are called *semolina* and are a "harder" starch than the first-break flour. Semolina flour is preferred for fine pasta. The large pieces of grain are fed through as many as four or five sets of rollers with semolinas sifted out after each break.

The semolinas and the first-break flour are then finely ground between smooth rollers and sieved to remove any lingering, flattened, pieces of the bran. While flour, traditionally associated with rich classes, but with the development of improved industrial milling of flour it became readily available to all. The germ and the bran are now often sold as animal feed. In addition to its widespread apeal, white flour has a longer shelf life than brown (whole wheat) flour because it lacks the embryo oils that tend to become rancid over time.

Graham flour, which contains the germ and the endosperm but not the bran, also becomes rancid quickly. Nevertheless, the whole wheat flour is much nutritious than the white flour. As white flour is lower in protein and fat, and lacks the fiber of the bran and the vitamins of whole flour. Most while flour sold in the United States is "enriched" with a mixtur of vitamin B1 (thiamin), niacin, and iron. With the modern increasing awareness of good nutrition, whole wheat or partialy refined flour has experienced an increase in popularity. Wheat "berries" (soaked, whole grains), sprouted wheat, cracked wheat, and wheat germ are all now frequently added to specialty breads.

Rye

In contrast to wheat and barley, which were both deliberately brought into cultivation, rye (*Secale cereale*) appears to have begun its association with man as a weed in wheat and barley fields. Instead of fighting the weed, people simply adopted it as a cultigen. As might be expected from this historical account, rye is native to Southwest Asia and commonly referred to as a secondary crop. Initial changes in rye paralleled those of the primary crops with which it grew. By about 3000 B.C., it appears to have been fully domesticated and its cultivation spread across Europe during the next 1000 years. Rye is hardly grain capable of germinating at 1°C above freezing and maturing when temperatures are as low as 12°C (55°F). Its roots can reach 2 m (over 6.5 feet) in length, allowing it to grow in even relatively dry habitats.

Rye has been called "poor man's wheat," because it has historically been grown in northern Europe in areas where wheat would not grow well. Black breads are widely used in Northern European regions which reflects the widespread cultivation of rye in these regions. Unlike wheat, the protein of rye is useless for making light bread. Consequently, even black bread usually contains wheat as well as rye flour. in the United States "rye" breads are made with half wheat and half rye flour. Rye is cultivated less than most of the major cereals (except triticale) discussed in this chapter while rye flour contains compratively large amounts of lysine. In addition to its use as a bread flour, rye is is consumed as forage and feed, planted for erosion control, and fermented in the production of rye whiskey and Dutch gin.

Triticale

Wheat and rye share not only a host-weed relationship, but also a common ancestor. This kinship is reflected by the fact that they can be crossed (albeit with some difficulty) to produce fertile offspring. While this potential for hybridization has been known for over 100 years, it is only recently that plant breeders have taken advantage of the fact and produced a fertile hybrid crop. The result of hybridization produced Triticosecaele or triticale, has been hailed as the first truly man-made cereal. Both hexaploid triticale, resulting from the doubling of the chromosome complement of hybrids between tetraploid wheat and diploid rye, and octoploids, resulting from the doubling of hybrids from hexaploid wheat and rye, are now being grown. You might ask why such a cross was attempted or why triticale has become a commercial crop. Like most crosses that have been found useful to humans, triticale combines desirable attributes to its parents and exceeds both of them in some characters. Triticale prouduces higher yields than either parent in marginal grain-growing regions, is nutritionally superior to wheat, and has the hardiness of rye. Although early reports claimed that triticale was extremely high in protein, they were based on poor data.

Oats

Like rye, oats are generally considered a secondary crop. In Ethiopia today, they are simply gathered and resown with the wheat. In most other areas, they are now cultivated alone. Of all

Fig. 15.4. An oat plant and inflorescence.

of the cereals native to the eastern Mediterranean, oats appear to have been the last domesticated, perhaps as late as 1000 B.C. The first of all the domestiction of oats probably occurred in Europe rather than in the Near East. Early cultivars were diploid or tetraploid, but by the first century A.D. the hexaploid, *Avena sativa*, became established as the major cultigen. Until recent times, oats were a major cereal of temperate areas. Part of this former prominence was due to the use of oats as feed, particularly for horses. Horses were introduced into Europe as draft animals about 200 B.C., but initially did poorly because thenative grasses did not provide good forage and, in the wet southern regions, horses suffered from foot problems.

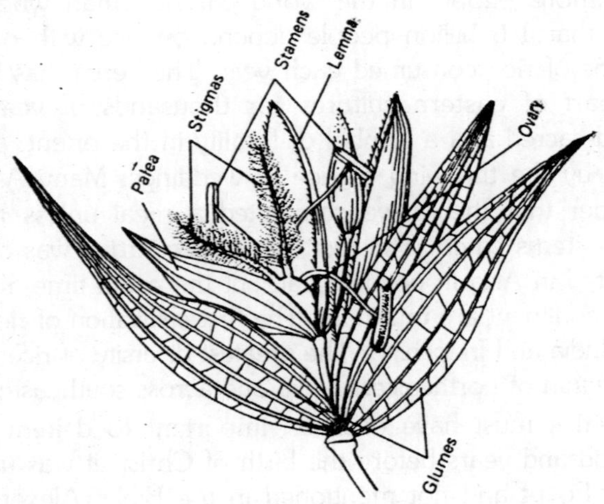

Fig. 15.5. Flower of an oat plant.

Only after the eighth century A.D. did several developments lead to adoption of the horse as the primary draft animal (replacing oxen). First, horseshoes, a protection against foot injuries and diseases began to be used. Second, a type of harness that did not choke horses whenthey were hitched to a plow or wagon was introduced from China. Horses were widely used. Peasants preferred horses to oxen because they were much faster than lumbering ungulates. The widespread use of horses necessitated a source of good feed, and oats served this purpose. The Romans took up cultivation of oats for animals but referred to the Germans as "oat-eating barbarians," reflecting the Roman opinion that the grain was unsuitable for human consumption. The protein and fat content in Oat are very high and provide straw that is good sources of roughage. Oat cultivation thus became an integral part of the former European agricultural system. In the United States as elsewhere, oats were extremely important until mechanized equipment obviated the need for farm draft animals after the 1930s and 1940s. Vestiges of the former widespread cultivation and the use of oats are evident in the American love of oatmeal and in the popular song of the 1940.

Rice

Rice (*Oryza sativa*) is the major agricultural crop in world. Wheat out ranks it in terms of the number of tons produced per year. but more people in the world eat rice than wheat. It is estimated that 1.6 billion people depend on rice with over 200 million tons of rice consumed each year. The cereal has been an integral part of eastern cultures for thousands of years. It is considered sacred and a symbol of fertility in the orient, a feeling expressed by the throwing of rice at weddings. Many Asians do not consider that they have completed a meal unless rice was among the items eaten. The fact that *Oryza sativa* was originally domesticated in Asia in unequivocal, but the exact time and place within the continent are uncertain. The first cultivation of rice began in China, India and Indochina. The greatest diversity of rice varieties is in the region of northeastern India and across southeastern Asia.

While rice must have been an important food item in Asia several thousand years before the birth of Christ, it was unknown in ancient Egypt and not mentioned in the Bible. Alexander the Great's conquests led to the first mention of the grain in Europe about 320 B.C. During the next 300 years before Christ, rice cultivation spread across the Middle East and into Egypt. But the fifteenth century, Spain and italy were both growing rice. The Portuguese introduced rice into Brazil and West Africa.

In the sixteenth century, the English started importinc rice from the Island of Madagascar off the east coast of Africa. One hundred years later (1965) one of these British ships, sailing home loaded with rice, was blown off course and landed in Charlestown, South Carolina. This unplanned visit led to the first plantings of rice in the United States, but commercially profitable operations began only much later, in 1912 in California. Australia, now a large rice producer, began commercial cultivation as late as 1925.

In the United States , rice constitutes a major crop in some areas of California. Arkansas. Texas. Louisiana and Mississippi. There are now a large number of different kinds of rice and numerous methods of rice cultivation. Rice is often divided into three subspecies. each of which produces a slightly different cooked produce. One of the subspecies. commonly called the *indica* type, is labeled in this country as long-grain rice. When cooked. the grains are relatively dry and separate easily from one another when "fluffed." In United States the *sativa* or *japonica* type is known as short-grain rice. After boiling, it becomes quite soft and often slightly gluey. The third type, *javanica*, is grown only on a very limited scale in the equatorial part of Indochina.

Rice jeeds large quantities of water in order to grow well, but it does not need to grow in standing water. Rice grown without standing water is known as *upland rice*. but upland rice will produce a harvestable crop only where there are 1.5 to 2 m of rain per year. The world's largest producer of upland rice is Brazil. Huge acreages of humid tropical rainforest in Brazil have been cleared and planted with rice. Unfortunarely. there are usually not enough nutrients in the soil to support rice cultivation formore than a year or two. After the rice potential has been depleted. fields are converted to pastures often seeded with African tropical grasses. Most of the rice is grown in standing water. It can grow partially submerged because the anatomy of stem is specialized and allows oxygen to reach the roots. We tend to call flooded fields of rice. "paddies," but the world "paddy" is also applied to unhusked rice grains.

Wet rice cultivation can permit almost continuous cultivation (assuming that temperatures are favourale). The standing water rots the plant material remaining after harvest and harboars algae that provide green fertilizer. Flooding the fields also prevents the growth of terrestrial weeds. In eastern countires, rice cultivation often involves extensive hand labour. The seed is sown in nursery beds and the seedlings individually transplanted into the fields for maturation. In the United States. seeding is often done by planes that broadcast the grains directly across the fields. Special harvesters have also been designed that mechanically harvest the crop when ripe. A third method of cultivation makes use of natural flood water. In floodplain areas, seeds are sown on the plain at times of low water. As the water rises, the rich grows apace with it. In some cases, the plants can reach 4 to 5 m in height before they finish growing. Harvesting is often done in such areas by boat.

With the arival of modern milling procedures, the world switched from brown to polished rice. The loss of the bran and germ was a subtle dietary change that had profound consequences. In the nineteenth century, a number of Japanese navy men experienced a loss of muscle tone in their arms and legs as a result of nerve inflamation.The disease, later diagnosed as beribert, soon broke out among thousands of people in Asia and Africa who subsisted on a rice diet. A Dutch doctor, Christian Eijkmann, working in Java was among those who concluded that the disease resulted forma dietary deficiency agravated by eating foods high in carbohydrates. It was later conformed that a lack of vitamin B1 was the main cause of the illness. The brans of grains are naturally high in vitamin B1, and brown rice is particularly rich, containing 0.40 mg per 100 g of grain. Polished rice contains only 0.04 mg per 100 g of grain. Several ways have been devised to get around the vitamin deficieny. In Japan, fish oils compensate to some degree.

The parboiling unmilled rice causes the outer layers of the endosperm to absorbe some of the vitamin before polishing. Rice can also be enriched by spraying it with synthetically produced vitamin B1 after polishing. Puffing rice was an American invention that uses enriched, partially milled rice. Rice is native to warm regions, but for many years, production in the wet tropics was unsuccessful because of lodging and/or because soils in such regions are poor and require large amounts of nitrogenous fertilizers. The development by the International Rice Institute in the Philipines of dwarf varieties such as IR8, which do not lodge, expnded cultivation in tropical wet areas. Newer varieties, such as IR36 developed by the Institute, combine the qualities of IR8 with increased pest resistance and a wide tolerance of poor soil conditions. The genus Oryza has large number of other species, most of which occur in Africa. One of these species, *O. glaberrima* domesticated in western Africa where it is still grown, althogh its cultivationis being replaced by that of *O. sativa*.

Wild Rice

Wild rice is not a member of these genus Oryza, but rather a New World species, *Zizania aquatica*. The species, like its relative

Texas wild rice, was used as a source of grain by New World indians, although it was never really domesticated by them. The inflorescences of wild rice shatter, and the grain has traditionally been collected by beating mature inflorescences while they were held over canoes. Native peoples roasted the grain, poured it into deer-skin-lined pits, and trampled it to remove the husks. In the United States the world rice cultivation began only in 1959 with the diking and flooding of areas of poor farmland in the northern part of the country. Plant breeders were finally successful in selecting a nonshattering strain. Nonshattering varieties raised yields from100 to 700 pounds per acre because they reduced losses incurred during harvesting of the older shattering types. Combines equiped with oversized wheels now harvest the grain. Over 17;000 acres are currently under wild rice cultiation, but the demand for the gourmet treat still exceeds the suply, and prices remain very high.

Sorghum

Sorghum (*sorghum bicolar*) is the best of all millets and is a grain native to Africa, but beyond this simple statement there is little agreement among botanists about its history. Some workers have favoured a single point of initial domestication within Africa and a diffusion of its cultivation outward. Others, such as Jack Harlan, postulate as many as three independent episodes of domestication. The date of this event or events is also speculative, with dates ranging from 2200 to 4000 B.C. In any case, since it was brought into cultivation, sorghum has been subjected to intensive divergent selection leading tonumerous, distinctive types. There has been much discussion and little agreement about the best method for grouping these types.

According to recent simple botanical solution suggests that the cultivated types should be grouped into 5 main biological race. Numerous intermediates between these races occur because of crossing practices. Nevertheless, cultivated sorghums are usually grouped by farmers into four main types based primarily on the use of the sorghum. These groups are (1) grain sorghums, (2) sweet sorghum or sorgo (used for animal feed), (3) Sudan grass (a different but related species), and (4) broomcorn or broom millet. The hybrid sorgos grown in thesouthern part of the United States are crosses between members of two of the biological races (kafir and durra). Sorghum a grain of hot regions with too little rainfall

(500 to 1,500 mm per year). The leaves are waxy and bicolour roll up to reduce the evaporation and thus aid in coping with water stress.

Sorghum is today the major grain consumed by many Africans and Indians. Dried grains are often ground and used to make flat breads. There are even pop sorghums similar to our popcorn. The grain is also used to make a beer. If sorghum is cut while in the milk stage (the same stage at which we pick sweet corn), it can be used for silage. Inthe United States, almost all of the sorghum grown is used for feed, although in some localities special syrup varieties are cultivated.

In order to get syrup the sorghum is harvested before flowering and its stalk crushed. The expressed juice is not purified, so the final, thick, sweet syrup resulting from the evaporation of most of the water retains a characteristic flavour. Broomcorn is a type of sorghum with an open inflorescence. The tillers are cut and stripped of grain and the stiff inflorescence "branches" used for making whisk brooms. There has recently been an increased interest in sorghum because of its drought tolerance and potential use on land that is a marginal for other grains. It is also a potentially useful crop for the making of alcohol.

Millets

Millets consists several grass. it does refer to a single grain. Numerous forage grasses and even sorghum are often called millet. Two of the species in this list. *Eleusine coracana* and *Pennisetum americanum*. are important compared to other millets as human food. The first of these. known most commonly as finger millet, was domesticated in north eastern tropical Africa. It is now a staple in the eastern and central parts of the continent. One of its important character is that it can be stored for long periods utp 10 years and more. Weevils avoid the grain and the grain itself does not deteriorate. In Africa, it is generaly ground inot flour andmade inot large, flat breads, or eaten as porridge. Once ground, however, the flour has to be consumed quickly because the germ is not removed and theflour rapidly becomes rancid. Finger millet can also be malted and used tomake beer.

Pennisetum americanum. The second species that constitutes an important human food also known as pearl, bullrush, spiked, or cattail millet. It is more drought-resistant than any of the cereals. Morevoer, it will give economically profitable yields on exhausted and nutrient-poor soil. Consequently, pearl millet has become the food upon which millions of Africans living on the edges of the Sahara now depend. Like finger millet, pearl miller stores well, but it is susceptible to weevil infestationand bird predation. It can also be ground intoflour, eaten like rice, or malted.

Corn

This is the most important cereal food in the New World in contact to the Old World. Maize formed the basis of all of the major new world civilizations, the Mayan, Aztec, and Inca, although *Amaranthus* species (Amaranthaceae) were very important in some regions in pre-Columbian times. The decline in the use of amaranths can be attributed in part to the fact that the early Spanish forbid their use because the Mayans used them in religious services as well as for food. Corn convert carbon dioxide and water into foodstuffs more efficiently than any other modern cereal grains. Moreover, it can be cultivated in both tropical and temperate areas. As children we learned that the Indians showed the Pilgrims how to grow corn and beans using fish as fertilizer. The beans and corn was grown together and this method spread throughout North and South America by the time Europeans reached the New World.

Presumaly, the practice developed without any real knowledge of themutual benefits that the system provides. An indian legend says the association came about when man-corn was looking for a wife. Squash asked to be considered, but she was rejeced because she had the habit of wandering haphazardly over the ground. Bean, onthe other hand, clasped the corn stalk so dearly that corn was assured of her fidelity, and consequently chose her for his bride. Whatever the origin of the biculture, the association proved to be highly successful in terms of both agronomy and nutrition. All of the legumes plant consumed as pulses have association with nitrogen-fixing bacteria which provide nitrogenous compound that the corn can use. The two kinds of food also complement one another in terms of human nutritional needs. Because of its importance, there has been great interest in the origin and subsequent evolutionary history of corn. It has been agreed for some time that corn was intially domesticated in the highlands of Mexico and that its cultivation spread north and southward. Charred cobs over 7,000 years old have been found in Mexico.

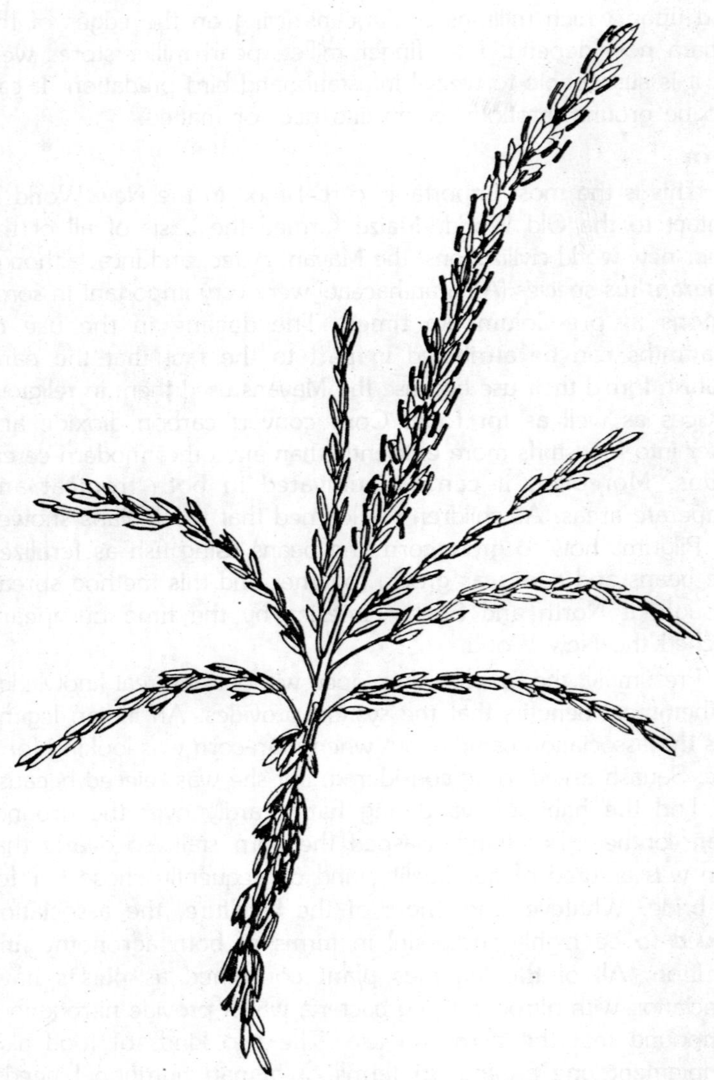

Fig. 15.6. A corn plant with an individual staminate flower.

The 7,000-year-old archaeological corn cobs excavated from dry, pre-Columbian caves in Tehuacan, Mexico, are small and similar in appearance to a kind of corn called Argentine pop corn that is still cultivated in some remote areas of Mexico. These corns have relatively lare, soft glumes around the small kernels.

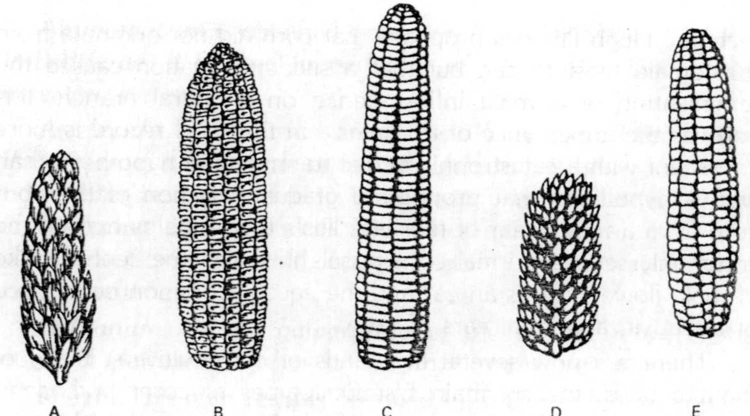

Fig. 15.7. The main corn varieties are: A—Pod corn; B—Dent corn; C—Flint corn; D—Popcorn; E—Flour corn.

Less than 2,000 years later, corn found in archaeological digs is very similar to modern corn. By the time of Columbus, corn was being transfered from Canada to Chile and about 300 major races had already been selected by native people. Still, there has been a century-long debate about the ancestry of modern cron and its relationship to one or both of two possible parents, *Tripsacum* and teosinte, both giant New World wild grasses.

One group of researchers, led by Paul Mangelsdorf and Edgar Anderson, produced a tripartite theory which postulated that *Zea mays* was a direct descendant of a wild South American form of corn that is now extinct. Teosinte, they suggested, was a "species" that resulted from hybridization between *Tripsacum* and corn once corn reached Central America. Modern corn itself resulted from subsequent introgression with *Tripsacum*. Other workers, including Harshberger, George Beadle, Hugh Iltis, and Waltor Galinat, have accumulated evidence that reflects corn must have been derived from teosinte. They initially placed teosinte in the same genus as corn, calling it *Zea mexicana*, and later in the same species as *Zea mays* subsp. *mexicana*. Many American and Mexican scientists have emphasized how easily the two taxa cross.

A cross between the two taxa made. Analysis of the progeny resulting from the crosses suggested to some that corn and teosinte differ by as few as 5 genes and that the conversion of a teosinte ear into a corn ear could have hapened relatively easily by converting the hard teosinte bracts into soft structures. Most recently, Hugh Iltis has proposed that corn did not originate from the female teosinte ear, but that a sudden mutation caused the feminization of a male inflorescence on a lateral branch. Iltis believes the appearance of true maize in the fossil record is more consistent with a catastrophic sexual tranmutation hypothesis than with a hypothesis that proposes a gradual selection of the corn corb from a female ear of teosinte. Iltis's theory helps explain the commonly seen abnormal corn tassels in which the central spike of male flowers forms an ear and the equally common occurrence of tassels in abnormal ears of corn.

There are now several thousands of corn cultivars, many of them hybrids, the six main historical types of corn: pod, dent, flint, top, flour, and sweet corn. In North America north of Mexico, two of these types, southern dents and northern flints (eight-row-corn) were most commonly grown at the time of European settlement. Columbus brought the corn back to Europe on his first voyage but did not gain much attention. Now every today, the popularity of corn is less in Europe than in other parts of the world, partly because most of Europe is too far north for corn to grow well. The European prejudice against corn was so strong that the Irish are said to have refused corn when drying of hunger following the decimation of the potato crop in the 1840s. Europeans insisted on grinding dried corn into cornmeal, which has very limited uses compared with wheat flour. While this was the same method of preparation practiced by the early European settlers in the New World (and used to make the classic cornbred, johnny cakes, etc.) native Aericans had more interesting and nutritionaly superior ways to use the grain.

The Aztecs are sweet maize porridge or boled the corn with a little charcoal (forming lime) to separate the grain coats from the endosperm. The endosperm was then mashed and the resultant dough formed around pieces of meat or mixture of beans, peppers, and meat. The mass was wrapped with leaves and streamed to produce tamales. Ground, dried cornmeal was used to thicken the Aztec's chocolate drink. Corn does not have the same nutritional value as other whole grains. Corn does not contains amino acids tryptophan and lysine, and it is relatively low in total protein. Since corn has no gluten, it can produce only flat breads such as tortillas or crumbly cakes. The nutritional deficiencies manifest themselves whereever corn has become the main, or the sole, part of the diet. Pellagra, a disease characterized by dermatitis. diarrhea. and dementia. was originally thought to be caused by a niacin (a member of the vitamin B complex) deficiency. When pellagra broke out among people in parts of Europe. northern Africa and the Americas who were existing on practically pure corn diets, scientists were puzzled. The pellagra broke due to deficiencies of niacin but corn does not have low levels of niacin.

It was subsequently discovered that tryptophan deficiencies were also a factor in causing pellagra. The diet of many rural Americans of molasses. corn bread. and salt pork made pallagra a serious problem in the south until 1914 when yeast was introduced as a food fupplement in this diet. Concerted efforts to improve dietary practices in this region led to a 74 percent reduction in pellagra in some southeastern states between 1928 and 1938. Corn crop grows in United States goes into animals. The entire plant can be chopped and fermented into silage.

Corn is also used for a wide variety of other purposes. Corn starch can be hydrolized to make corn syrup. which contains glucose and fructose used in baby formulas and intravenous feeding. Purified. ground corn endosperm yields corn starch used to thicken cooking liquids and even to powder baby bottoms. Maize beer. or chicha. is produced by fermenting hydrolyzed corn starch. Traditionally. corn kernels were chewed before fermenting by Central and South American Indians in order to introduce salivary amylase to break down the starch.

Corn is an important adjunct in modern brewing operations and corn is a major ingredient in the productionof bourbon, and in the production of industrial alcohol. The cytology and genetics of corn can be easily and better understood than that of any other plant species. Since corn is monoecious plant so hybridization is relatively easy and breeders are able to produce superior hybrids. Plants to be hybridized are planted in alternate rows. The rows that serve as the female parent (those with ears to be used for seed) are detasseled by workers riding through the fields. Self-pollination is thus prevented. and all the seeds produced are of hybrid origin. In the late 1960s it was discovered that a male-sterile gene could be bred into corn lines. Using male-sterle plants, detasseling became unnecessary.

At first. the use of male-sterile lines saved money. but it soon became dramatically clear that the cytoplasmic sterility gene carried with it a susceptibility to a new race of the southern corn blight (*Helminthosporium maydis*, Dematiaceae). Once this fungus started infesting fields, it spread rapidly. In 1970, warm, moist weather encouraged fungal growth, leading to devastating crop losses amounting to over $4 billion in the United States. Luckily, the next season was dry, averting a second major outbreak. New immune hybrid crosses without this particular male-sterility gene were redistributed during the next few years. This episode startled plant breeders into realizing the need for genetic reserves of this and other major crop species.

Forage Grasses

It would be difficult to explain all the forage grasses used throughout the world due to lacks of statistics, because these are usually consumed on the farm and thus do not enter into trade. it is estimated that in the United States, forage crops (which includes forage legumes as well as forage grasses), cover five times the acreage of all cereal grain crops combined. Land that is not well suited for high-intensity agriculture or that is "resting' between periods of cultivation is often allowed to become grassland.

In addition to renewing the humus in the soil, grasses help prevent erosion and increase soil drainage. In order to maximize thefood value of forage land and the improvement of the soil, grasses are normally sown with legumes. Forage can be grazed directly, made into hay, or used for silage. The dehydration of green forage produced hay. Which contain moisture content of 15% water or less. The quality of hay is determined by the plants that are incorporated into the hay, the amount of leaf material relative to stem material. the time the forage was harvested,and the amount of weathering and handling to which it has been subjected. The form in which the hay is presented to animals (loose, pelletized, etc.) also has an effect on its suitability for different animals.

The cellulose content in the grasses are very high and for most animal the cellulose is digested via bacterial fermentation before the hay can be utilized. In addition to reducing cellulose to simpler compounds that can be digested by vertebrates, the intestinal bacteria produce necessary amino acids and vitamins. Ruminants have large gastrointestinal tracts that contain abundant bacteria. The rumen, or primary area where bacterial fermentation occurs, is at the beginning of the digestive tract of ruminants.

Consequently, hay with a high percentage of grass relative to legumes provides adequate feed for ruminants.

In nonruminants, or animals with simple stomachs, fermentation takes place near the end of the digestive tract. Consequently, these animals cannot take full advantage of bacterial fermentation and their feed must contain proportionately higher quantities of proteinaceous foods. Since over 95 percent of the hay produced in this country is for ruminants, most hay has a high proportion of grasses relative to legumes. The controlled anaerobic fermentation of undried forage, or stalks of corn, sorghum or cane produce silage. This process of making of silage is called *ensiling* the place where silage is made, a *silo*. Silage can be classified into high moisture (over 70 percent water), wilted (Between 60 and 70 percent moisture), or low moisture (40 to 60 percent water), but the basic process involved in the production of all three types is the same. The cut plant material is loaded into a silo. For a few minutes after the plants are cut, the multiplication of aerobic bacteria takes place on the plant surface. Bacteria used carbohydrates exuded by the cut plants for its multiplication.

Both heat and carbon dioxide are produced by the respiration of these bacteria. Once these microorganisms have exhausted the oxygen supply, a second kind of bacteria that can grow without oxygen begin to multiply. These bacteria produce latic acid andsome additional organic acids. The acid levels in the silo eventually reach 8 to 9 percent. This concentration of acid eventually "pickles" the plant matter and keeps it from rotting. If the acid content is high enough and air is excluded, silage can be kept for several years. Many of the major forage grasses of the United States were introduced from Europe or Africa, although native species are also used. Usually, forage grasses are perennials, but some such as corn, sorghum, millet, and Sudan grass are annuals.

16

LEGUMES

Next to grass family plant of legume family are quite important to humans. Every major civilization since the development of agriculture has had a legume as well as a grain as part of its support system. Rice and soyabeans, barley and lentils, and corn and bearns are classical grain-legume combinations that not only taste good, but make good nutritional sense as well. Legumes are members of the Fabaceae (*Leguminosae*), or bean, family. The family is large and diverse, containing perhaps as many as 12,000 species that botanists usually groups into three subfamilies on the basis of their floral characteristics. Although the flowers are different, they have same kind of fruit called legume which is formed by a single carpel and split along two opposite longitudinal margins at maturity to release its seeds. Its common name is bean pod. However, some members of the family have fruits so modified that they only slightly resemble a common bean.

Of the three subfamilies, the Faboideae is the most importatn as a source of fod crops. Members of the other subfamilies are occasionally used for food, but more commonly to improve soils, as forage, for their gums as source of dyestuffs or as fuel. Here we deal primarily with the pulses (dried legume seeds used for human food) and forage crops, but we also include tamarind and carob, two other foods obtained from the Caesalpinoideae. The tendency of roots to form association with bacteria, primarily members of the genus *Rhizobium* (Azobacteriaceae) is an important characteristic of most species in the Faboideae, and some species of other subfamilies as well. These bacteria infect the roots of

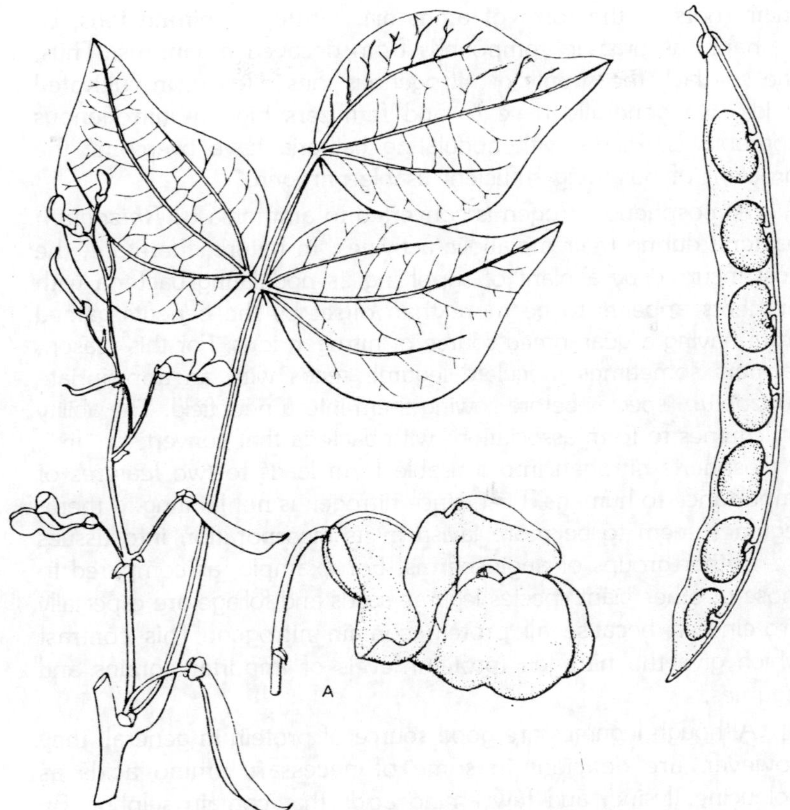

Fig. 16.1. The common bean, like most members of the Faboideae, has irregular flowers with the upper petal or standard exterior to the wing petals. The keel, formed by the joining of the two lower petals, houses the style and variously joined stamens.

these legumes, causing the production of swollen areas called *nodules*. The bacteria live within these nodules and absorb nutrients from the host plant. It has been shown by growing plants which normally have root nodules in conditins free of *Rhizobium* (but with abundant fertilizer) that the bacteria use resources that the legume host could put into its own growth.

Under natural conditions, however, abundant fertilizer is not present. in particular, one component of fertilizer, nitrogen, is often in short supply. Although nitrogen gas is the most abundant substance in the earth's atmosphere, flowering plants cannot use elemental nitrogen. Instead, they must absorbe nitrogen through their roots in the form of ammonia, nitrite, or nitrate ions, or perhaps. as organic compounds from decayed organisms. Thus, the soil lack the supply of nitrogenous ions. Hence, in cultivated fields. we generally have to add fertilizers high in nitrogenous compounds. Plants with nodulating bacteria have overcome the problem of obtaining sufficient usable nitrogen.

Atmospheric nitrogen is converted to ammonia by *Rhizobium* bacteria during their metabolic activities. In nature, therefore, the cost incurred by a plant of supplying its nodulating bacteria with nutrients appears to be more than offset by the benefits gained from having a guaranteed source of nitrogen ions. For this reason, farmers sometimes inoculate legume seeds with an appropriate *Rhizobium* species before sowing them into a new field. The ability of legumes to form associations with bacteria that convert, or "fix", atmospheric nitrogen into a usable form leads to two features of importance to humans. First, since nitrogen is not limiting to them, legumes seem to be more lavish in its incorporation into tissues than other groups of angiosperms. For example, as compared to those of other plant species legume seeds and foliage are especially protein-rich because all protein contain nitrogen. This contrast which give the nitrogen (protein) levels of important grains and legumes.

Although legumes are good source of protein in general, they however. are deficient in some of necessary amino acids as isoleucine, lysine. and few amino acids that contain sulphur. By eating other kinds of plants containing the amino acids that are deficient in legumes, it is possible to obtain a complete spectrum of amino acids necessary for human nutrition. The amino acid profiles of several pulses discussed in this chapter compared to beef steak, and examples of the use of complementary amino acid sources to provide all the protein components necessary in the human diet. The second feature of legumes of particular value to humans is the fact that the legume-bacteria association usually produces an excess of usable nitrogen in the soil. This means, in effect, than legumes fertilize the soil besides providing a food crop. Becaue of this property of legume crops, farmers often rotate (alternate) them with other crops that would normally require the application of nitrogenous fertilizers.

Legumes are replanted after growing nonlegume crops for 1 or 2 years as the nitrogen ion content in soil get depleted.

While we could have included many of the legume species used by man, we have chosen first to discuss 10 of the most important pulses and 2 caesalpinoid fruit crops. We then look at several of the major legume forages grown in the world today. In our discussion of the importatn pulses, we begin with native Old World species and finish with American contributions to the world's supply of these protein-rich plant foods.

PULSES

Lentils

Lentils are among ancient plants known to have been cultivated by man. Carbnized lentils found in Neolithic villages in the Middle East have been dated as being between 8 and 9 thousand years old. Analysis of the fossilized seeds indicates a period of domestication had occured years before the time when

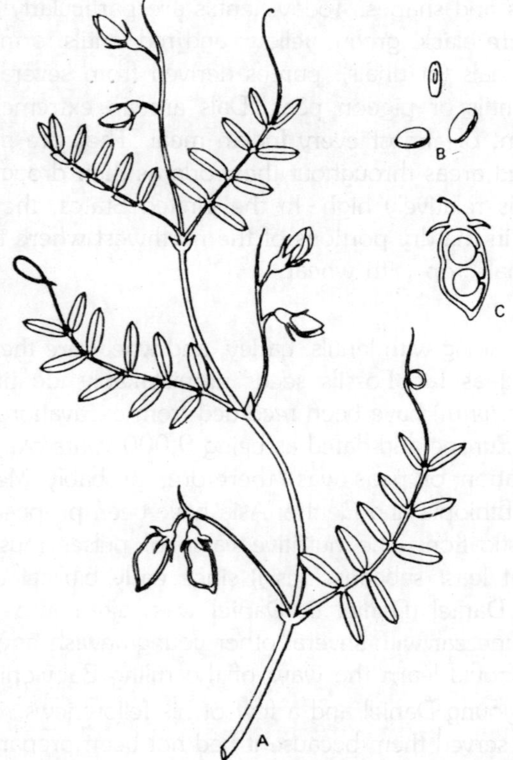

Fig. 16.2. Lentils. A—A branch; B—Seeds; and C—A seed pod in section.

the seeds were deposited. Fossil excavations has discovered these seeds and found them to be larger than the seeds of wild lentils when still grow in the area. After initial cultivation of the crop in the Middle East, lentil use began to spread around the Mediterranean. By 2200 B.C., lentils began to appear in Egyptian tombs. Lentils are the first pulse to be mentioned in the Bible. According to the story in Genesis, Esau, the first-born twin of Rebecca and Isaac, sold his birthright to his brother, Jacob for a meal of red lentils. Flattened shape of the seed similar to the human eye lens is responsible for its common name as scientists name *Lens culinaris*. One to three seeds are produced per pod, a smaller number than produced by most other pulses. However, lentils rank fifth among the major legumes in proetin content, and they are among the most digestible of all the commonly eaten pulses.

Over the many thousands of years during which they have been cultivated, lentils have been selected by humans for a range of colours and shapes. Today, lentils are particularly important in India where black, green, yellow, and red lentils form the basis of the most dals (or dhals), purees derived from several pulses, but usually lentils or pigeon peas. Dals are an extremely important component of almost every Indian meal. They are mainly grown in semiarid areas throughout the world as their drought resistance capacity is relatively high. In the United States, they are grown primarily in the dry portions of the northwest where they serve as a rotational crop with wheat.

Peas

Peas, along with lentils, barley, and wheat are the oldest foods discovered as far. Fossils seeds unmistakably identified as peas (*Pisum sativum*) have been recorded from excavations in the Near East and Europe and dated as being 9,000 years old. The original domestication of peas was, therefore, probably Mediterranean, although Ethiopia and Central Asia have been proposed as centers of domestication. The nutritive value of pulses must have been known (at least subconsciously) since early biblical times. In the Book of Daniel (Daniel I), Daniel was taken into the court of Nebuchadnezzar with several other young Jewish boys so that the Israelites could learn the ways of the ruling Babylonians.

The young Daniel and a few of his fellow jews refused to eat the meat served them because it had not been prepared according to Herbrew specifications and the eating of it would have been a violation of their dietary laws. Daniel asked that they be allowed to eat pulses (probably peas) rather than meat. The story continues that, despite the caretakers' fears that the boys would show the effects of this "poor" diet (and hence bring punishment from the king), they granted the request. However, when they were finally brought for an audience at the royal court, Daniel and the other youths who had refused to eat the king's meat had fatter and more beautiful faces than those who had forsaken their religion. During the Middle Ages in Europe, dried peas were the mainstay of the peasant population. The familiar nursery rhyme: *Pease porridge hot, Pease porridge cold, Pease porridge in the pot, Nine days old*...refers to a thick broth or steaming pudding made from dried peas that, if we are to believe the text of the rhyme, could be served from the same pot for over a week. Peas were not eaten as a fresh green vegetable until the seventeenth century, when a Dutch horticulturalist bred and offered in trade the first "garden" varieties. Green peas were enthusiastically accepted by the members of the court of Kind Louis XIV, even though, at this early time, fresh peas would probably have been much inferiour to our modern fresh peas.

Peas bear the fourth rank after soyabeans, common beans and peanuts and are still one of the most important pulse in world. Breeding in recent times has been for more pods per node so that the fruit crop will mature synchronously, facilitating mechanical harvesting. Chinese snow peas and the new sugar snap peas are pea varieties eaten when the pods are still immature and tender.

Broad Beans

Today, broad, or fava, veans (*Vicia faba*) are associated with the Mediterranean region believed to the original home of the species. Cultivation was widespread in this region in prehistoric times, and writings attest to the cultivation of broad beans by Egyptians, Greeks, and Romans. From southern Europe, broad beans spread to Asia (China is now the world's largest produer, and the New World. While *broad beans* are primarily a crop of cool regions, it was brought to South America from Spain in 1543 by the second governor of Colombia. In North America, Canada is a larger producer than the United States. In the United States, production has so declined in recent years that most of the canned and dried beans offered in supermarkets are imported. The seeds

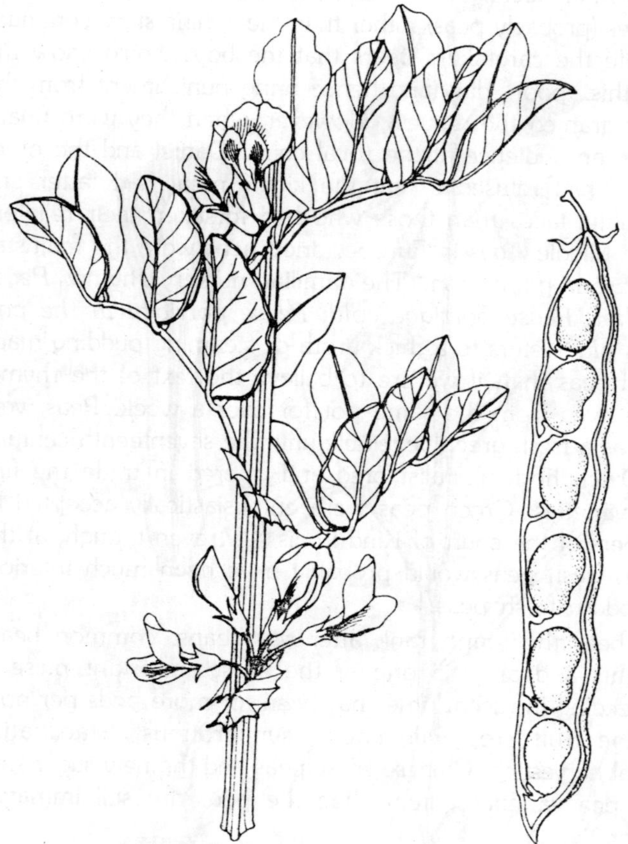

Fig. 16.3. Foliage, flowers, and pod of a broad bean.

of broad beans are said to produce favism, a disease involving hemolytic anemia.

The illness, resulting from the breakdown of red blood cells, is most commonly found in people of Mediterranean origin. Now studies have shown that disease is due to lack of enzyme glucose-6-phosphate dehydrogenase which is a genetic disorder. The disease is no longer thought to be caused by beans seeds. When oxidative agents such as the fava bean alkaloids are ingested by individuals with this disorder, the anemia is aggravated. Oddly enough, before the advent of modern medicine, the disease was often advantageous to individuals who had to because it provided a resistance to malaria in much the same way as sickle-cell anemia conferred malaria resistance to blacks in Africa.

Chick-peas

East of the Mediterranean, another legume replaces peas and broad beans and becomes, along with lentils, a basic component of the diet of literally millions of people. Chick-peas (*Cicer arietinum*) have a record of cultivation that is only slightly younger than that of peas adn lentils. The first certain records, from Turkey, are dated to be 7,400 years old. The area of origin of the speceis is assumed to be in northeast Africa. By 200 B.C., chick-peas were introduced to India, which now produces 79 percent of the world'd crop.

Cultivation also spread around the Mediterranean, and today chick-peas form a part of the cuisines of Italy, Spain, Morocco, and Algeria. Brought to Mexico by the Spanish, who call them garbanzos, they now constitute, along with beans, a Mexican nonmeat source of protein.

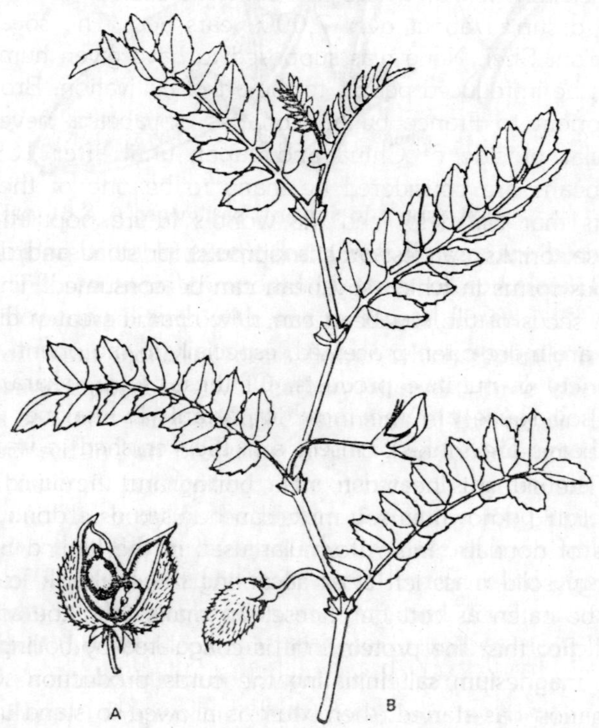

Fig. 16.4. Chick-peas have pods which contain two or three seeds each. A—Pod; B—Plant.

Soyabeans

Soyabean is the best of all the legumes being considered as poor man's meat in China. This indicates its importance in this populous country. While all of the legumes we are discussing can be considered as partial substitues for animal protein, the soyabean has more protein and greater versatiilty, than any of the others. Soyabeans also have less carbohydrate per unit weight than any of the other pulses except peanuts. It has fairly high concentration of amino acids especially the sulphur-containing compounds in contrast to others. All these properties make soyabeans to be crowned the king of legumes in the world.

In the western hemisphere, soyabeans are used primarily for oil extraction and animal food. In the east, the use of the beans is quite a different story. Soyabeans are thought to be native to northeastern china, where the domestrication is hypothesized to have occurred about over 7,000 years ago. The soyabean was the plant Shen Nung was supposed to have given human beings when he introduced people to the art of cultivation. Brought from the orient to France by missionaries, soyabeans never became popular outside of China and Japan until after 1890. Now, soyabeans are considered by many to be one of the principal plants that will help feed the world's future population. Curd, cheese, drinks, sauces, greens (sprouts) for salad and oils are the various forms in which soyabean can be consumed. The dreid or fresh seeds of the soyabean can, of course, be eaten directly, but they are more often processed, especially in the orient, to provide a variety of nutritive products. Of these, tofru, okara, soy milk, and soy sauce are the most important. In making tofu, dried soyabeans are soaked, rinsed, and then crushed in water.

The slurry is heated to near boiling and the liquid decanted. This liquid, known as soy milk, can be used as a drink and is the basis of nondairy infant formulas used in the United States. The spongy solid mass left after decanting the soy milk, called okara can be eaten as cottage cheese. To make tofu, the soya milk is used. For this, the protein in it is coagulated by boiling the liquid with magnesium salt initiating the curds production. Once curd formation has started, themixture is allowed to stand until almost all of the protein has coagulated. In a manner similar to that of fresh cheese making, the curds are scooped from the now thin liquid and drained as much as possible. After sufficient draining, the curds settle into firm, smooth-textured tofu cakes. Tofu is extremely nutritious, very digestible, and quite bland, qualities that allow it to be used in a variety of dishes. Soy sauce is another product made from the beans.

In making the sauce, okara, sometimes mixed with wheat, is formed into cakes. The cakes are placed in a cool, dark environment for 3 or more months, during which they become encrusted with fungi. After removing the crusted part of fungi cakes are then soaked in a salty water. The filtered, now flavoured salt water is soy sauce. Before you recoil from the thought of pouring juice from moldy soyabean paste on your food, remember that most cheeses are made by allowing fungi to act on milk curds (in Roquefort, blue Danish, and Stilton cheeses, the blue veins of fungi are still visible). In fact the fermented cakes, after they have been soaked and rinsed, are eaten in China like cheese. However, above method of soaking the mermented soy cakes in salt water is not employed to make American soya sauce. Rather the salt water is flavoured for this purpose. One of the conspicuous nutrients that most beans lack in vitamin C. The Chinese, so dependent on soyabeans, have managed to overcome this shortcoming by sprouting the beans.

Soy sprouts contain abundant vitamic C and are eaten in salads. It is little wonder that soyabeans have been held in such high esteem in the orient for so long and are now coming into their own in west as well. Soyabeans were not commercially planted in the United States until 1920. Since then, a meteoric rise in soyabean productionhas made the United States the world's largest soyabean supplier and led to the legume's being called a "Cinderella crop." Before the twentieth century, soyabeans were not used as livestock food because the raw vegetable contains a trypsin inhibitor. Trypsin is an enzyme that is necessary for animal digestion of prootein.

In the early part of this century, it was discovered that heating destroyed the inhibitor. As a result of this finding, soyabeans became a major component of animal feeds, and worldwide demand for the crop soared. Henry Ford is the first American who took interest in soyabean, he thought that they had as much potential as a raw material for manufactured goods as they had as a source of protein. The Ford Motor Company built three factories in the 1930s where soyabean oil was extracted and used in the production of paints and plastics. Ford himself are soyabeans at every meal and had a suit made from "soy fabric" that he wore to a convention. His company also sponsored a 16-course soyabean dinner at the 1934 "Century of Progress" show in Chicago. But still a little of huge American crop is consumed directly.

Over half of production is exported, and most of the rest is fed to animals, processed into oil, plastic, and adhesives. Spinning soy proteins into *texturized vetetable protein* (TVP) provides a mean to overcome this problem. Soy protein, processed in this way can be flavoured to imitate various meats and utilized as a meat substitute or added as a filler to canned or processed meats. Many dry dog foods produced in America contain processed soyabeans as well as grain by-procucts.

Pigeon Peas

While the name pigeon pea (*Cajanus cajan*) might not seem like a familiar one, you have probably eaten them without realizing it. In the United States, pigeon peas are often used as the basis of split pea soup (although true peas are also used). In other parts of the world, however, they constitute a common and important part of the diet. There is dispute about the place of origin of the crop. Due to prominence of India, most botanists consider India to be the place of its origin. Other botanists suggest that the pea was first cultivated in northern Africa because there are 4,000 year-old fossils thought to be pigeon peas from this regior.. Whatever the origin the legume is now of major importance in India, which produces 95 percent of the world's crop.

Pigeon pea proves to be a good crop for farmers in utilizing the marginal agricultural land, because it can grow well in poor soil. In India, the dried peas are used like lentils in the preparation of dals. Since pigeon peas have remained a crop grown on a small scale, and one of poor farmers, few large-scale efforts have been made to produce new varieties and better yields.

Black-Eyed Peas (Cowpeas)

Afria (Ethiopia) is usually considered to be the original area of domestication of black-eyed peas, *Vigna unguiculata*. There is still an extraordinary amount of diversity in the peas from this region. Spread of their cultivation seems to have occurred both westward within Africa and eastward to India. The majority of the crop, however, is still grown in Africa where seeds, sprouts, and

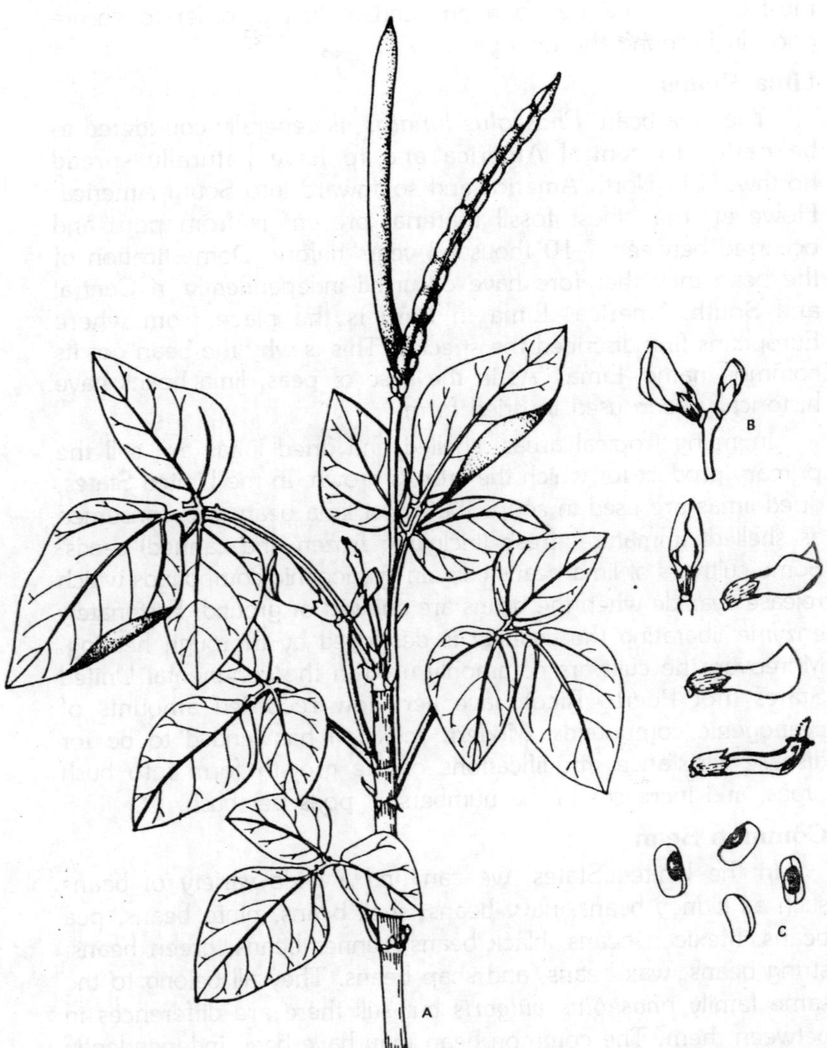

Fig. 16.5. Black-eyed peas. A—Plant: B—Flowers in profile and section; C—Shelled peas.

leaves are eaten. Slaves brought from western Africa introduced the *Black-eyed peas* to the United States. Today, they are part of the regional cooking of the American south. In part of the southern United States, hoppin' John, a mixture of rice, black-eyed peas (or cowpeas as they are often called), and salt pork is the traditional dish of New Year's Day. Supersition says the stew must be eaten before noon on January first in order to ensure good luck during the new year.

Lima Beans

The lima bean, *Phaseolus lunatus*, is generally considered to be native to central America and to have naturally spread northward in North America and southward into South America. However, the oldest fossil material present is from peru and occurred between 7-10 thousand years before. Domestication of the bean may therefore have occurred independently in Central and South America. Lima in Peru is the place from where Europeans first discribed the species. This is why the bean got its common name 'Lima'. As in the case of peas, lima beans have historically been used in dried form.

In many tropical areas, shelled and dried limas are still the primary product for which the crop is grown. In the United States, dried limas are used in soups, but most lima beans are consumed as shelled, immature, fresh (including frozen and canned) seeds. Some cultivars of lima beans contain cyanogenic compounds which release cyanide when the beans are chewed or ground. Fortunately enzyme liberating the cyanide is destroyed by thorough heating. Moreover, the cultivars commonly used in the continental United States (not Puerto Rico) have very low recorded amounts of cyanogenic compounds. Modern selection has tended to be for disease resistance, modifications of the growth form into bush types, and increases in the numbers of pods per bush.

Common Bean

In the United States, we can find a wide variety of beans such as kidney beans, navy beans, shell beans, pinto beans, pea beans, Mexican beans, black beans, runner beans, green beans, string beans, wax beans, and snap beans. They all belong to the same family *phaseolus vulgaris* but still there are differences in between them. The common bean may have been independently domesticated in both Central and South America. Early dates of fossil cultivated material from Central America are about 7,000 years ago. Recently uncovered material from Peru is as old as, or older than, the Mexican material. By the time of European arrival in the New World, common beans were an important dietary item for native peoples throughout North, Central, and South America.

Today, it is the most widely cultivated species of *Phaseolus*, and the second most important legume in the world (next to soyabeans) in terms of tons produed per year. Because plants of the common bean are naturally vines (Bush forms are recent selections), they were historically often grown with another crop such as corn to provide the vines with a support. Because of its association with nitrogen-fixing bacteria, this mixed cropping was also advantageous for the corn. Native Americans, without realizing the reason, must have found that growing these two crops together increased production of both. The Indian dish succotash reflected the method of mixed agriculture and was made from dried corn and dried common beans.

Our modern idea of succotash as being mixture of corn and lima beans comes from the name on packages of frozen vegetables. Refried beans, mashed precooked beans fried in oil, commonly associated with Mexican cooking today, came into existence only after the Spanish had introduced domesticated animals to Central America. The Aztecs appear not to have had much cooking oil, probably because Central American wild game had little fat, and native oil seed sources were rare in Central America. Moreover, outside of the llama, dog, guinea pig, and turkey, New World Indians had no domesticated animals and thus no ready source of lard. Frying, therefore, was presumably an innovation brought from Europe and very successfully used with this native American legume.

Peanuts

Peanuts are rather considered as nuts or as an oil seed crop than as a pulse but like any other grain legume seeds are cooked in various parts of world. The species from which we obtain peanuts, *Arachis hypogaea*, is native to central South America, perhaps eastern Peru. Domestication probably occurred first in southern Bolivia and northwestern Argentina. By the time Columbus reached the New World, peanuts were cultivated throughout the warm regions of the Americas. The Portuguese took peanuts to Africa where their cultivation was quickly adopted.

In west African countries it constitute an important dietary item. Peanuts were also taken to southeast Asia via the Philippines by the Spanish. Peanuts are called by different names in various parts of the world. The British name groundnut, or ground pea, refers to the way in which peanuts bear their fruits. Like other legumes of the subfamily Faboideae, peanuts bear pea-type flowers above the ground. Self-pollination occurs within the flowers. Crop

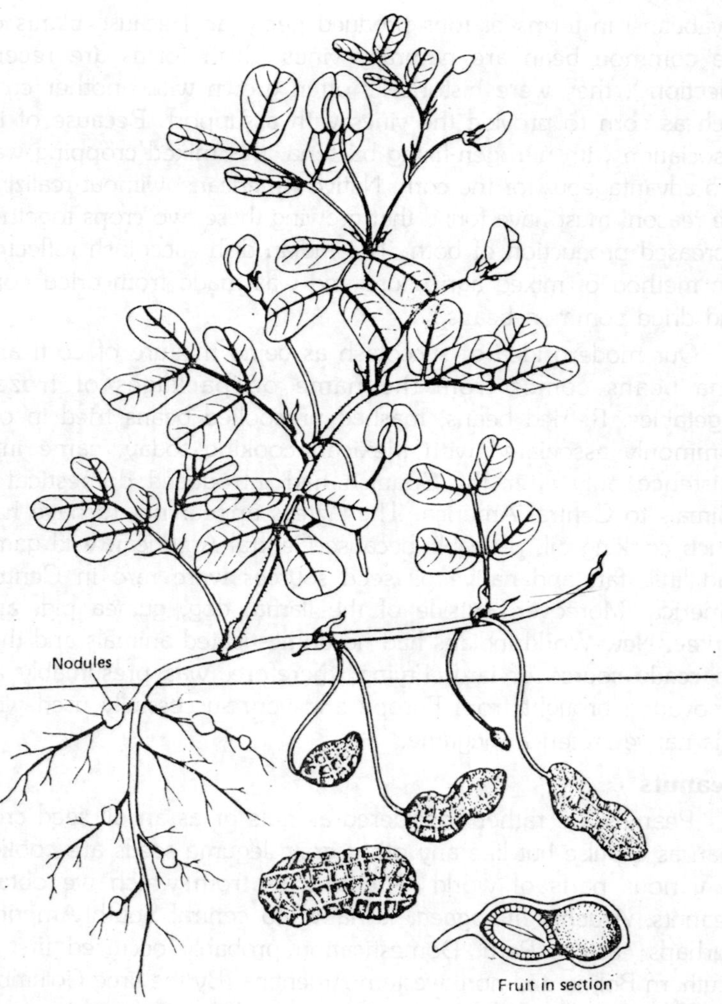

Fig. 16.6. After fertilization, the flower pedicles of the peanut curve downward, and the developing fruit is forced into the ground by the proliferation and elongation by cells under the ovary. The legume subsequently develops underground.

of the legume is not produced on stem, instead the pedicels or flower stalks curve downward after fertilization. The cells under the ovary begin to divide, producing a "peg" that forces the ovary into the ground. As the peg develops a cap of cells forms next to the withered style. This cap protects the ovary as it is pushed into the soil in much the same way as the root cap protects a root.

Downward growth stops as soon as the expanding ovary reached a few centimeters into the soil. The ovary turns sideward and matures underground.

Many people unfamiliar with the taxonomy or flowering process of peanuts think the fruits are roots or tubers. Peanuts are also called goobers, a name that was brought with the peanut back to the New World by African slaves. Crop's popularity in black Africa and its secondary introduction to North America by slaves of southern plantation is responsible for its widespread production in American southeast. It is not due its being native in New World.

With the decline of the cototn empire, peanuts have partially replaced cotton as a major crop in that part of the country. The popularity of peanuts now extends to all parts of the American population. Most of the peanut consumption in the United States today is in the form of peanut butter, although, a substantial part of the crop is used for hog feed, South American Indians ground peanuts and produced a produc similar to modern peanut butter, but "development" of the creamy paste now used is often attributed to a physician, thought by some to be John Harvey Kellogg. The doctor, a health food advocate, was searching for an easily digestible, highly nutritious food that required little effort (chewing) to eat.

In 1890, salted peanut paste was being produced. The name Kellogg is probably known by most people as the name of a modern corporation that has gone beyond the sphere of health foods. The Kellogg Company was founded by W. K. Kellogg, John Harvey's brother. The two brothers worked together, and the company originally manufactured flaked cereals that had been developed in the Battle Creek Sanatorium run by John Harvey Kellogg. Due to their high protein content, ability to grow in poor soil and in tropical region they have been subjected to much agricultural research. Yet they are susceptible to infestation by a fungus which secretes aflatoxins, chemicals deadly poisonous to humans. Deaths from eating contaminated peanuts have been greatly reduced by instituting proper methods of handling and storage which can virtually eliminate the danger of infestation.

Tamaaind and Carob

Tamarinds (*Tamarindus indica*) have been used in Africa, where wild plants grow in the tropical dry savannas, and in southern Asia for thousands of years. The ripe, long frown pods

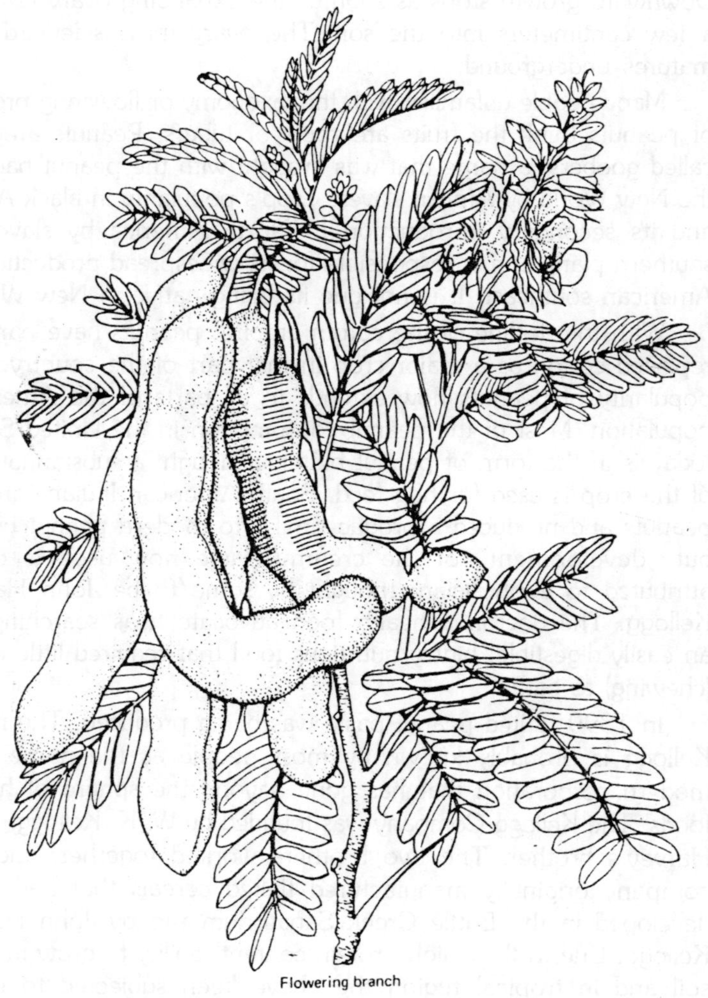

Fig. 16.7. Tamarind pods hanging from a branch.

on tall, spreading trees are iused primarily for their tart, rather sticky pulp, but the seeds can also be roasted or boiled and eaten after the removal of the seed coat, Tamarinds are rarely grown in plantations, plants are usually grown from seed and pods harvested from semiwild trees. In the United States and Mexico, tamarinds are used primarily as a flavouring in sauces. In India, chutneys and sauces are manufactured from the fruit pulp. Carob is anoher caesalpinoid legume that produces pods used primarily for their pulp.

The species (*Ceratonia siliqua*) is native to the Mediterranean region. Its common name St. John's bread comes from the fact that it constituted the "locust" on which John the Baptist fed. Their extreme uniformity in size made them useful as weights for small quantity of precious substances such as gold, in ancient times. Our modern unit the "carat" used for gold and jewels is a reflection of this former use. Traditionally, carbo pods were gathered from wild trees and the sweet pulp chewed is the mesocarp. The seeds have also been used to make a cofee-like beverage.

Today, carob trees are propagated by seed or by grafting. Since the species is primarily dioecious, grafting assures a large proportion of female trees. Production is still highest in the Mediterranean region, with the tiney island of Cyprus the world's largest producer. Carbo is mostly used as a substitute for chocolate in united states and as a source of locust gum, a polymer extracted from the endosperm of the seeds.

FORAGES

Just as legumes complement grains for human consumptin, legume forages supplement forage grasses and feed grains for domesticated hervibores. Legume grass herbage association not only provide best mixture of food for grazing animals. It is beneficial for other reasons also. Research has shown that legumes alone do not build soil as well as herbaceous mixtures. One of the best ways to improve soil is to plant a grass-legume mixture and use the area for prudent pasturage. Simply harvesting the hay from such a field removes much of the potentially available nitrogen and the plowing under of the cover without cutting precludes any direct profit.

Grazing animals on the field allows conversion of part of the vegetable biomass into a cash crop (meat), permits the roots of the grass and legumes to build up the soil, and provides the addition of considerable nitrogen in the form of animal urea. This method can prove to be the better one than letting the field go fallow or a simple crop rotation system if conditions are favourable. The growth of grass with legumes also circumvents problesm presented by pure fields of forage legumes. Ruminants do poorly on a diet consisting principally of legumes because fermentation of large amounts of fresh leguminous material in the rumen cases bloat.

Forage legumes also generally do not grow well in pure stadns because they cannot compete successfully with weedy grasses and forbs. The growing of legumes with desirabel grasses alleviates the problem of weed invasion. Direct baling of the quality of hay desired can be achieved by growing a suitable mixture of grass and legumes, if the crop is to be cut for hay only. Several hundred legumes are used for forage on a worldwide basis, but a few predominate, especially in this country. Unlike grasses, native legumes are used les for forage in the United States than introduced, domesticated species. Some temperate forage legumes which are important in the United States. Of these, alfalfa and the clovers are by far the most important.

We therefore discuss these two kinds of forage in some detail and then mention some of less important Americna legume forages. Alfalfa (*Medicago sativa*), or lucerne as it is called in most of the Old World, is king of the forage legues. Alfalfa was probably the only forage crop cultivated in prehistoric times. It is now extensively grown on every continent except Antarctica, and it has an area of cultivation that exceeds 33 million hectares. This acerage is a **litte** larger in size than the amount of land to which the common bean is planted (Ca. 24 million hectares in 1981). Although cultivated alfalfa is tetraploid, primitive diploids occur in Iran and other parts of the Near East believed to be the area of original domestication.

It appears likely that the spread of alfalfa cultivation followed the adoptin of horse husbandry, extending to the north, east, and west from the Near East after 1300 B.C. It was introduced to North America about 1850, although it had been cultivated earlier in Mexico and South America. An initial stumbling block to the use of alfalfa in north temperate areas was its lack of winter hardiness. After 1900, breeding efforts overcame this problem, and the acreage devoted to alfalfa in the northern United States and Eurasia escalated rapidly. Alfalfa is a pernnial that is grown from seed. It can be used as a pasture crop harvested and used in hay. In warm regions with an adequate moisture supply, as many as six to nine cuttings can be made per year. While making an excellent fooder in dried form pure alfalfa hay is comparatively expensive.

Consequently, alfalfa is usually grown with forage grasses so that the bailed harest will contain a proper mixture of the two kinds of forages. True clovers, species of the genus *Trifoliu*, are the next most important group of forage legumes. In the United

States, red clover (*Trifolium pratense*) and white clover (*T. trepens*) are the most widely grown. Both are perennials and both are native to the Old World. The two clovers were independently introduced to North America by European colonists. Red clover is the most extensively grown of all of the clovers on a worldwide basis, but white clover is the most important in many temperate regions.

Like alfalfa, clovers are often planted with grasses and can be used as pasturage or cut for hay. The other three major clovers grown in the United States are alsike clover (*T. hybridum*), arrowhead clover (*T. vesiculosum*), and crimson clover (*T. incarnatum*). Sweet clovers, primarily *Melilotus officinalis* and *M. alba*, are the other important "clovers" grown in the United States. One superior ability of sweet clovers to improve soils makes them very popular. Both species produce long roots that aerate the soil well during growth. Once senescence begins, their roots rapidly decay, releasing precious nutrients at considerable depths in the soil.

The other major group of pasture and hay legumes frown in the United States are the lespedezas (*Lespedeza* spp.). All are native to eastern Asia and were successfully introduced into this country only after 1919. Although it started late, there is a rapid proliferation of their cultivation. Annual lespedezas make excellent pasture forage, and plants can be harvested for hay at the end of the summer. A last forage legume that has become increasingly popular is bird's-foot trefoil, *Lotus corniculatus*. This species in extremely adaptable to a wide range of climate and soil conditions and it is very persistent. It has proved to be especially useful in the northeastern United States, where heavy soils predominate.

17

Fats and Oils

Vegetables fats and oils, also called *fixed oils*, are chemically esters of fatty acids and a trihydric alcohol, glycerol. Waxes are esters of fatty acid and a monohydric alcohol. Fats are solid or semisolid at ordinary temperature whereas oils are usually in a liquid state. The fats contain stearic or palmitic acid and the oils contain oleic acid. They are actually colourless and tasteless but traces of essential oils or pigments may impart colour and taste to them. If stored for long and at high temperatures they break into various aldehydes, ketones, etc. and develop rancid taste and smell. Edible fats and oils are so much in demand these days that the governments of various countries are paying great attention in finding out new sources of vegetables oils. Fats and oils are classified into four types:

1. *Semi-drying oils*. They form soft, solid films only after longer exposure to air. They include cottonseed oil, corn oil, sesame oil, sunflower oil, mustard oil etc.
2. *Drying oils*. They harden into elastic films on exposure to air. They include linseed oil, tung oil, walnut oil, perilla oil, soyabean oil, safflower oil, etc.
3. *Non-drying oils*. They do not form solid films on exposure to air. They include olive oil, castor oil, groundnut oil, etc.
4. *Vegetable fats*. Including coconut oil, palm oil, coca butter, nutmeg butter etc. they are solid or semi-solid at ordinary temperature.

The coconut oil, soyabean oil, cottonseed oil, sesame oil, mustard oil, and groundnut oil are more wel known edible oils

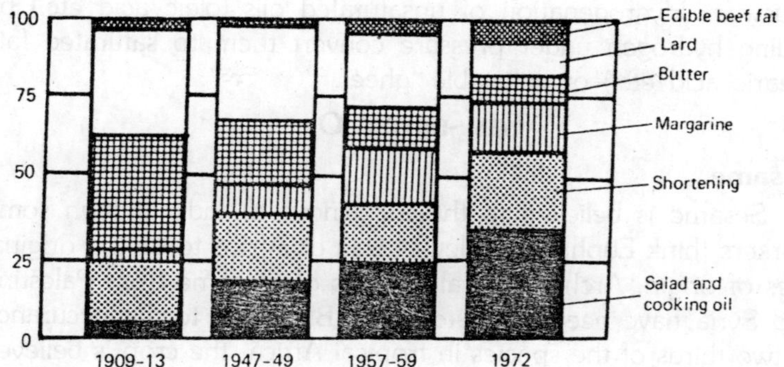

Fig. 17.1. During this century, there has been an increase in the consumption of fats and oils.

and are widely used as a cooking medium. Some of the oils are used for massaging the body and hairs. Soaps and candles are extensively manufactured from oils. They are also used in preparing paints, varnishes, furniture and leather polishes, oil papers, putty, laxatives, lubricants, cosmetics, waterproof fabrics, inks, artificial leather illuminants and oil cloth etc.

Extraction

The fatty oils are insoluble in water and are present in the form of insoluble droplets or deposits within cells. Most oils are stored in seeds, but they may also be found in fruits, stems, tubers, etc. Much of energy needed for germination of seeds is provided by the oils stored in them. Extraction of oil involves separation of kernels from the seed coats, hot or cold expression, solvent or screw extraction and centrifuging. In hot expression the separated kernels are first cooked in steam to facilitate oil flow during extraction.

In cold expression the kernels are directly crushed into a fine meal without any previous treatment. The oils are then removed from the meal by organic solvents in the case of non-edible oils but in case of edible oils removed by screw or hydraulic pressure. The solvent evaporates at 60° to 113°C and the extracted oil is left behind. The pressure breaks the cell walls and the fat droplets are extracted. The oil is then usually run over filter presses. Further purification may be performed by heating the filtered oil so as to saponify free fatty acids. The unsaturated edible 'oils' are now being converted into saturated 'fats' by the 'vanaspati' manufacturers since the latter is preferred very much by the modern houswife. Hydrogenation of unsaturated oils (oleic acid etc.) by adding hydrogen under pressure convert them to saturated fats (stearic acid etc.) or vegetable 'ghee'.

SEMI-DRYING OILS

Sesame

Sesame is believed to the indigenous to India though some workers think Euphrates valley of Iraq or Africa to be the original sites of origin. Archaeological findings of seasame from Palestine and Syria have been dated to 3000 B.C. Due to the occurance of two thirds of the species in tropical Africa, the crop is believed to be African. On the other hand no wild species have been found in West Asia.

India is the largest sesame-growing country. It is also an important crop in Burma. The crop is grown in India from ancient times. Though most of the workers do not believe in its Indian origin. *Nayar* and *Mehta* (1970) believe it to have originated in Ethiopia or peninsular India or in both independently. Some carbonized seeds have been found from the ancient site of Harappa, where Indus valley civilization flourished about 5,500 years ago. Domestication of Sesame in the remote past can be indicated by the common use of 'til' seeds in the religious rites of Hindus. It is also believed to have been under cultivation in China from the ancient times.

Sesame growing regions

The chief sesame-growing countries lie in the topics as well as in the temperates. They are China, India, Burma, African (Sudan, Nigeria, Uganda) and Latin American countries (particularly, Mexico). Sesame oil is a very importatn vegetable oil in China, India and Mexico. In India seasame is cultivated in Uttar Pradesh, Madhya Pradesh, Rajasthan, Maharashtra, Andhra Pradesh and Tamil Nadu.

Botanical description

Sesame belongs to the family *Pedaliaceae*. The oil is also known as benne and gingelly. *Sesamum* has about 36 species. The plant is an annual herb. The leaves bear 1 to 3 axillary flowers. The fruit is a capsule. Dehiscence is by means of apical pores. Having white, brown or black appearance, the seeds contain about 50% oil (oleic and linolic glyce, rides), 18 to 25% carbohydrates and 20 to 24% protein.

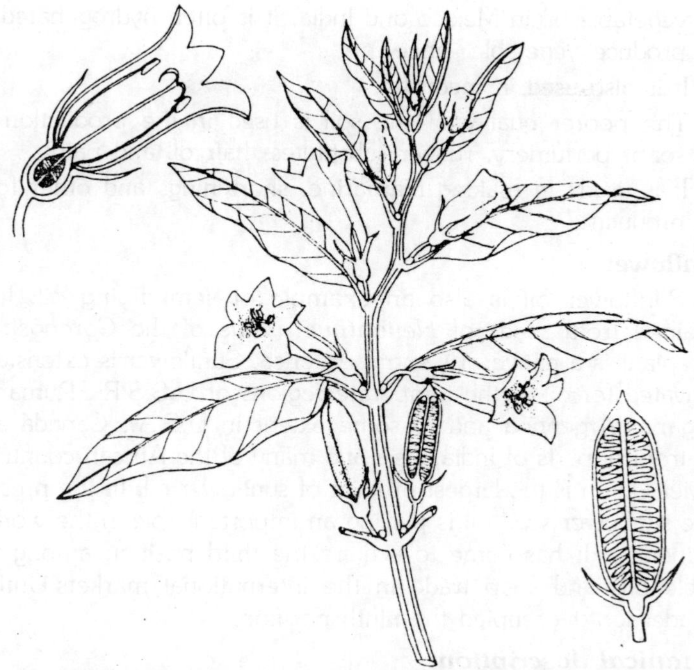

Fig. 17.2. Flowers of the seasame plant, usually grown for seed production, are sometimes cultivated in this country as ornamentals. A cross section of a flower (above, left), a seasame branch (middle) and a fruit (right) are shown.

Cultivation

It grows well in warm climates of the tropics and subtropics. It requires plenty of water. sandy soil is good for its cultivation. The seeds are sown broadcast or in rows. The plants are harvested by hand. The plants are cut and dried in heaps. Shaking the cut plant or beating or trampling upon the capsules are the methods employed to remove seeds from capsules. The sesame oil is extracted by cold pressure. The oil is of *semi-drying* type. The presence of *sesamin* and *sesamolin*, which are powerful anti-oxidants does not allow it to turn rancid.

Uses of sesame oil

1. In Africa sesame forms a part of diet in the form of porridge, soup and sweets.
2. The seeds are used in the confectionery and baking industires, 'Laddoo', 'rewri', 'gazak' and cakes are prepared from the seeds.

3. Sesame oil is used as a cooking medium. It is the principal vegetable oil in Mexico and India. It is often hydrogenated to produce vegetable 'ghee'.
4. It is also used in medicine.
5. The poorer quality of the soil is used in the production of soap, perfumery, rubber substitutes, hair oil and paints.
6. It is used in making margarine, shortenings and other food products.

Sunflower

Sunflower oil is also an example of semi-drying oil. It is obtained from seeds of *Helianthus annus* of the *Compositae*. The plant is a native of Central America. Sunflower is extensively cultivated for oil in the temperate regions of U.S.S.R., Rumania, Bulgaria, Argentina and to some extent in U.S.A., Canada and the tropical parts of India, Asia and many of the African countries. Soviet Union is the largest product of sunflower oil. In the present time sunflower seed oil is playing an important role in the world's oil indusry. It has come to acquire the third position among the edible oils and soap trade in the international markets.Until a decade ago it occupied the ninth position.

Botanical description

The plant is a coarse, tall, annual. It has a woody stem which is covered with rough hairs. It is of 5 to 10 feet in height. Leaves are large, ovate or cordate, dark green and are borne, in alternate manner. The inflorescence is capitulum or head which is usually terminal. There are two kinds of florets—tubular disc florets and ligulate ray florets in the in inflorescence. The seeds contain 55 to 60 percent of linoleic acid and 25 to 30 per cent of oleic acid. In cool temperate climates the seeds contain 70 per cent linoleic acid and 15 per cent oleic acid, while in hot tropics the linoleic acid is as little as 20 per cent and the oleic is as much as 65 per cent.

The semi-drying oil is usually 32 to 45 per cent. It is pale yellow in colour. The oil is extracted by hot or cold expression. Sunflower seed oil meal contains over 40 per cent high grade protein. It is sweet in taste and qualitywise is as good as sesame and soyabean meal.Sunflower oil after refinement gets a pleasant flavour and has excellent keepint quality.

Sunflower seeds unlike soyabean and copra do not require sophisticated machinery for processing. It is possible to grow three

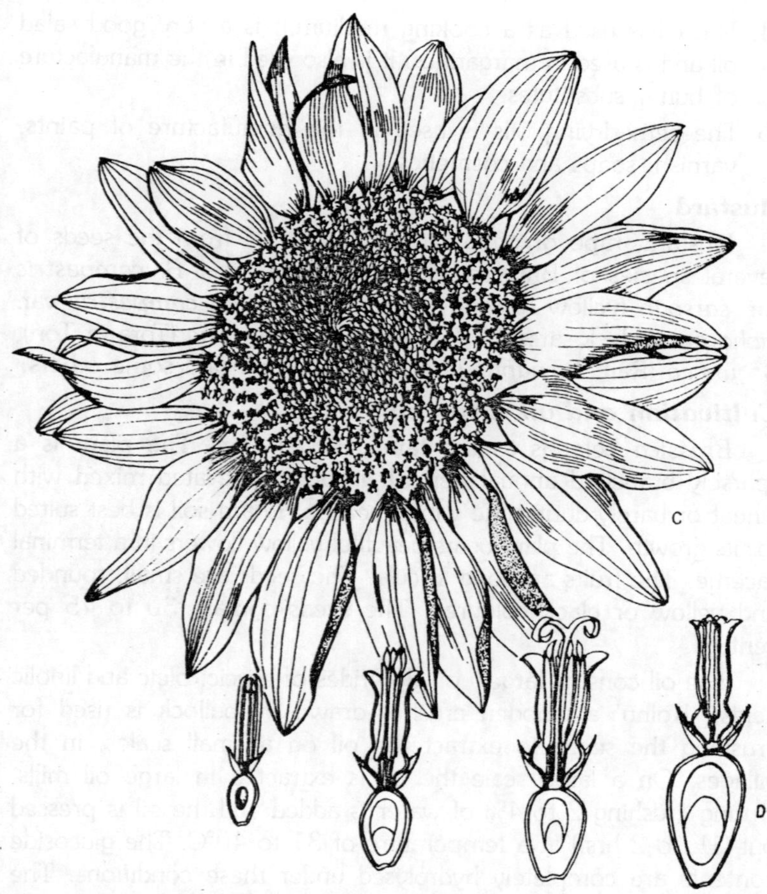

Fig. 17.3. Sunflower, the sequence of blomming and fruit maturation of a single flower.

sunflowr crops a year with an average yield of 20 to 25 quintals per hectare per season. Sunflower crop can be helpful in solving the problem of non-utilization of land in off season by adjusting it in any crop rotation. This is due to its day neutral native. This necessity has been increasingly felt with the extension of irrigation facilities.

Uses of sunflower oil

1. Sunflower yield honey and wax.
2. The seeds are a good poultry food.
3. The cake is used as a stock feed.

4. The oil is used as a cooking medium.It is a very good salad oil and is used in margarines. It is also used in the manufacture of butter substitutes.
5. The semi-drying oils is used in the manufacture of paints, varnish, soaps and cosmetics.

Mustard

Mustard, rape or colza oils are produced from the seeds of several species of Brassica. Some of them are : *B. campestris* var. *sarsoan* (yellow sarsoan or Indian colza), *B. campestris* var. *dichotoma* (black sarsoan), *B. campestris* var. *toria* (Tora or Tori), *B. juncea* (Rai), *B. napus* (the rape), *B. rapa* and some others.

Cultivation and oil extraction

Brassica belongs to the family *Cruciferae*. The plant is a sparsely branched annual herb. It is often cultivated mixed with wheat or barely during the winter season. Loam soil is best suited for its growth. The plant bears beautiful yellow flowers in a terminal raceme. The fruits are long silicula. The seeds are small, rounded and yellow or black coloured. The seeds contain 30 to 45 per cent oil.

The oil consists largely or glycerides of erucic, oleic and linolic acids. 'Kolhu' a wooden crusher drawn by bullock is used for crushing the seeds to extract the oil on a small scales, in the villages. On a large scale the oil is extracted in large oil mills. During crushing 2 to 4% of water is added and the oil is pressed out (1½ to 2 hrs.) at a temperature of 31 to 40°C. The glucoside contents are completely hydrolysed under these conditions. The crude oil is yellow, light brown or greenish in colour with a characteristic sharp taste and pungent smell.

Crude oil, when purified with sulphuric acid or caustic soda gets converted into a pale yellow oil with high viscosity and having properties midway between those of *semi-drying* and *non-drying* oils. BARC has developed TM-1 as a result of mutation breeding of mustard. It has a higher oil content than the popular Rai-5.

Uses of mustard

1. Mustard seeds particularly rai are used as spice in vegetable and pickle preparations.
2. The oil cake known as 'Khali' is very nutritious and is used as a cattle feed.

3. Mustard oil is edible and is used for cooking vegetables is northern part of our country. It is used for frying a variety of food preparations.
4. The oil is used for massaging the body and as hair oil.
5. With camphor it is used for giving relief from muscular rheumatism and stiff neck, etc.
6. The oil is also used as a fuel in lamps etc.
7. Mustard oil is used as a lubricant in machineries, etc.
8. The oil is used in the manufacture of soap and rubber substitutes.

DRYING OILS

Linseed (Flax; Vern. Alsi)

Linseed or flax is a source of an oil and a fibre. According to *Vavilov* the smal seeded types of linseed originated in southwestern

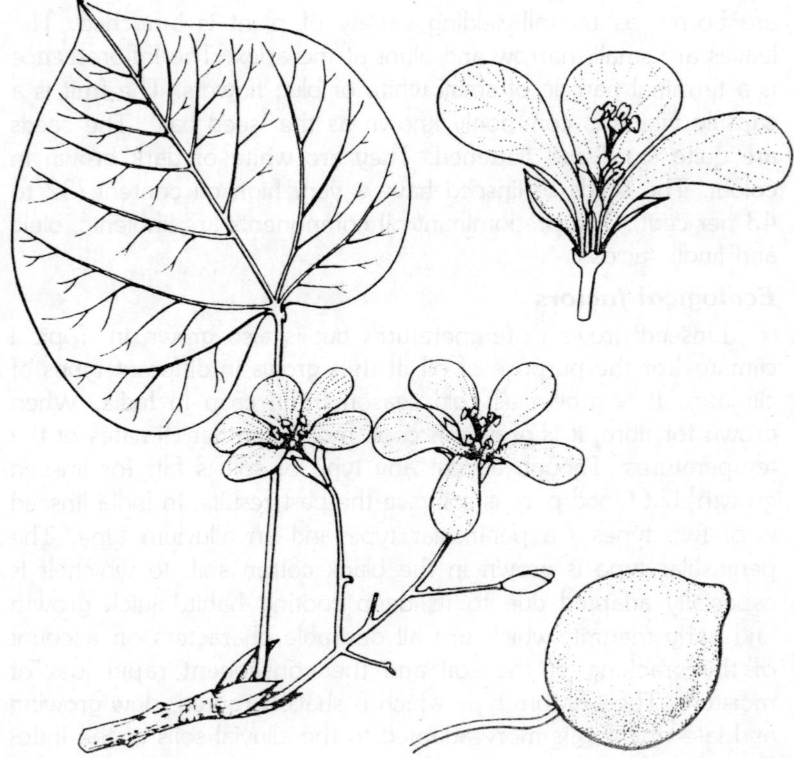

Fig. 17.4. A flowering branch of Aleurites fordii, the species that provides the seeds most commonly used for the extraction of tung oil.

Asia (Afghanistan. etc.). It spread northward to Europe and other parts of Asia and southward to India. The northern forms evolved inot early races with long unbranched stems whereas the southern forms evolved into late races with lot of branching in the stem.

Linseed gowing regions

Linseed as a source of fibre has been in use of more than 4,000 years. It is not extensively gronw in Canada, United States, Uruguay, Argentina, U.S.S.R., India and China. Linseed is chiefly grown in Bihar, Uttar Pradesh, Madhya Pradesh and Andhra Pradesh in India. It is also cultivated in Bengal, Punjab, and Maharashtra.

Botanical description

Belonging to the family *Linaceae*, linseed is a herbaceous annual herb reaching a height of 1 to 4 feet. The Indian linseed is of a shorter height of about 2½ feet. Large number of seeds are borne, as the oil-yielding variety of plant is branched. The leaves are small, narrow and blunt at the apex. The inflorescence is a terminal panicle of small white or blue flowers. The fruit is a capsule which is commonly known as the 'seed ball'. The seeds are quite small and flattened. They are white or dark brown in colour. The seeds of linseed have a very high oil content (32 to 43 per cent). The predominant oil components are linolenic. oleic and linolic acids.

Ecological factors

Linseed grows in temperatures but is also grown in tropical climates for the purpose of oil. It thus grows in different types of climates. It is grown as cold season (*rabi*) crop in India. When grown for fibre, it is grown in cool, moist summer climates of the temperatures. Though almost any type of soil is fair for linseed growth, but Good porous soil give the best results. In India linseed is of two types : a peninsular type and an alluvium type. The peninsular type is grown in the black cotton soil, to which it is especially adapted due to its deep rooting habit, quick growth and early maturity,which are all desirable characters on account of the cracking of the soil and the consequent rapid loss of moisture. The alluvium type which is shallow-rooted, slow growing and late ripening is more adapted to the alluvial soils of the Indo-Gangetic plains.

Cultivation

A fine, firm and porous seed bed is prepared by ploughing the land several times, followed by harrowing. The land is then rolled or planked. Usually, a drill or a tube attached to a plough through which seeds are dropped are used for sowing seeds. Sometimes the seeds are sown broadcast. It is sometimes sown as a mixed crop with gram or 'rai' (*Brassica juncea*). When sown alone in lines, the lines are a foot apart. The crop is usually harvested in 3 to 4 months, time after sowing.

Harvesting is done by hand as well as by machine. The plants are either uprooted or are cut by sickles. They are then taken to the threshing floor where the seeds are beaten out and winnowed. The seeds are cleaned and crushed to a fine meal. Oil is then extracted by heating or by means of solvents. To extract edible oils the cold expression is practiced. The oil is heated to 125°C to increase its drying property. Linseed oil is a very important *drying oil*. It is one of the oldest of oils and has been used as a drying oil for centuries. Linseed oil is yellow-brown. The liquid oil can be kept almost indefinitely in closed containers, but on exposure to air they become hard and form elastic films.

Uses of linseed oils

1. The oil is used as fuel and in leather production.
2. The press cake is used as stock feed provided the hot pressure process has been used. The cold dressed cake is poisonous due to the presence of a cyanogenetic glucoside, Limarin. The heating prevents the evolution of prussic acid from the glucoside and so it becomes edible.
3. The oil is chiefly used in the manufacture of paints, varnishes, linoleum, oil cloth, soft soap and printing ink.
4. The refined and unheated oil is used for cooking purposes in U.S.S.R. and other countries of Eastern Europe.

Safflower (Carthamus tinctorius)

Sufflower oil is a drying oil obtained from the seeds of *Carthamus tinctorius* of the family *Composite*. The plant is also well known as a source of an orange-red dye. It is native to India. It is widely cultivated in India, the West Asian countries, Egypt and many other east African countries, China and to some extent in U.S.A. and southern Eurpoe. Safflower seeds have been found in some 3,500 years old Egyptain tombs.

Botanical description

The plant is an annual herb. It grows very rapidly. It is highly branched. It grows to a height of 1 to 4 feet. Leaves are crowded near the base of the stem in the form of a rosette. Leaves are usually arranged in a spiral manner and are borne at irregular intervals. Tip of branches bear the leaves that are borne very closely thus forming involucral bracts of the inflorescence. The 'head' inflorescence is formed at the tip of the each branch of the plant. All the florets of the inflorescence are regular and tubular. The fruit is an achene. The seed contains 24 to 36 per cent of a very good drying oil. It does not dry as rapidly as linseed oil.

Uses of safflower oil

1. Decorticated (without fibrous seed coat) seed cake is a good stock feed (20 to 55% protein).
2. In India the oil is mainly used for food and cooking.
3. In other countries it is chiefly used in the paint and varnish industry.

NON-DRYING OIL

Castor

The castor oil is a very important industrial non-drying oil, obtained from *R. communis* seeds. Castor belongs to the family *Euphorbiaceae*. The plant is probably native of Africa. Now the plant grows in wild as well as cultivated state in the whole world in both the temperate and tropical regions. It is chiefly cultivated in Brazil, India, Mexico, U.S.S.R. and Manchuria. The plant is a coarse and erect herbaceous annual shrub.

The plant grows to a height of several feet. The leaves are typically large and palmately lobed. The flowers are very small and fruits are spiny. The fruits contain large mottled seeds containing 35 to 55 per cent oil. Cold expression is used to extract the oil from seeds. Oil may be further extracted from the press cake by means of organic solvents. The oil is thick, colourless or greenish.

Uses of castor

1. Sulfonated castor oils are used as dyeing aid and in finishing fabrics and leather.
2. It's water resistant quality makes it useful in number of articles. It is used for preserving leather.

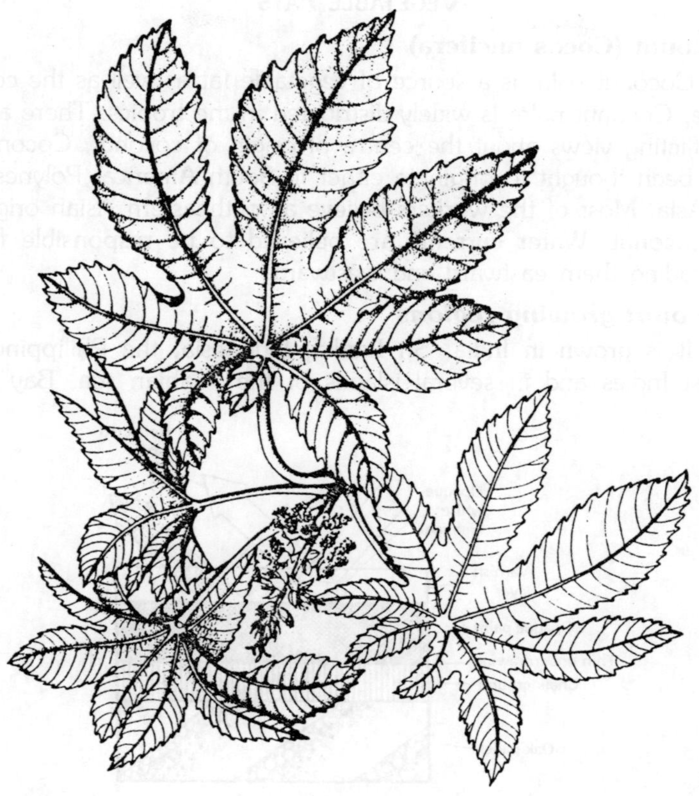

Fig. 17.5. Caster bean plant.

3. Paper pulp and cellulose are manufactured from stem.
4. Castor oil becomes a *drying* oil if it is heated and dehydrated. It can then be mixed with more expensive oils in the paint industry.
5. Acting as purgative, it is of medicinal use.
6. The oil cake is poisonous but is a good fertilizer.
7. The oil has a number of industrial uses. It is used in the manufacture of soaps, inks, plastics, imitation leather, linoleum, oil cloth, nylon, etc.
8. The oil is also used as a fuel in lamps, etc.
9. The oil used for lubrication. As a lubricant it is used in a number of machineries including that of aeroplane.

Vegetable Fats

Cocount (Cocos nucifera)

Coconut palm is a source of vegetable fat as well as the coir fibre. Coconut palm is widely distributed in the tropics. There are conflicting views about the centre of origin of coconut. Coconut has been thought to originate either in South America, Polynesia or Asia. Most of the workers believe a southeastern Asian origin of coconut. Water currents are believed to be responsible for spreading them eastward and westward.

Cocount growing regions

It is grown in India, Sri Lanka, Indonesia, the Philippines, West Indies and in several islands of the Arabian sea, Bay of

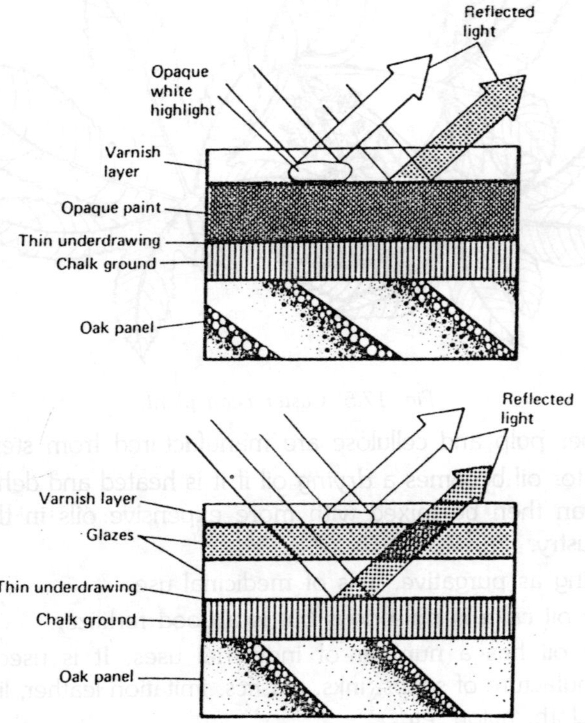

Fig. 17.6. Flemish artists added a new dimension to painting when they developed a technique of mixing pigments with vegetable oil, usually walnut or linseed, to produce translucent glazes. As is shown in the lower diagram, light penetrates the glaze layer and is reflected to create an illusion of depth that was not possible with previous opaque paints.

Bengal and Pacific Ocean. The Philippines is the largest product of coconut. India stands second in coconut production. In India cocount is grown extensively in the coastal regions of southern and eastern India. Tamil Nadu, Kerala, Maharashtra, Karnataka, Gujarat and Bengal.

Botanical description

Cocos nucifera belongs to the family *Palmaceae*. It is tall plant and attains a height of 80 ft. or more and lives up to 90 years but the dwarf varieties are small and live up to 30 to 35 years only. It has an unbranched stout trunk which has rings of leaf scars. Swollen base of the trunk is surrounded by a mass of unbranched roots. The turnk bears at the tip a crown of very

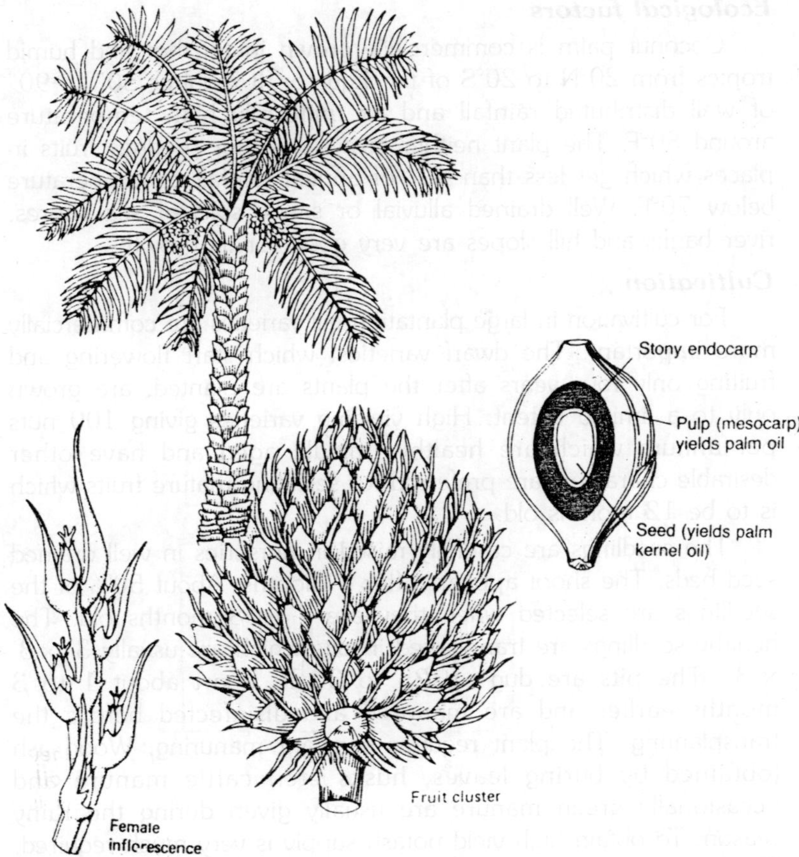

Fig. 17.7. Palm oil is extracted from the fleshy mesocarps of the palm fruits.

large pinnate leaves. The leaves are 6-8 ft. long and the leaflets are 2-3 ft. long. Numerous male and few female flowers are born in spadix inflorescences in the axils of leaves when the plant is 8-10 years old. Fruits are single seeded, three angled and are borne in clusters at the top of the plant. They may be 6"–12" in length. The colour of the fruit varies from dark green to deep orange or brick red. The fruit is a drupe. The exocarp and mesocarp are thick and fibrous (source of coir) whereas the endocarp (shell) is a hard structure. There are three pores at one end of the shell 'eyes'. The embryo lies below one of these pores. Internally there is thick albuminous endosperm (meat) attached all over the endocarp. In young fruits there is a large cavity filled with watery fluid and enclosed by the endocarp.

Ecological factors

Coconut palm is commercially grown in the hot and humid tropics from 20°N to 20°S of the Equator. It requires 50" to 90" of well distributed rainfall and an uniformly high temperature around 80°F. The plant neither grows well nor produes fruits in places which get less than 40" of annual rainfall or temperature below 70°F. Well drained alluvial or sandy soils of sea shores, river banks and hill slopes are very good for its growth.

Cultivation

For cultivation in large plantation tall varieties are commercially more important. The dwarf varieties, which start flowering and fruiting only four years after the plants are planted, are grown only to a limited extent. High yielding varieties giving 100 nuts per annum which are healthy, middle aged and have other desirable characters are preferred for selecting mature fruits which is to be 12 months old.

The seedlings are carefully raised in nurseries in well drained seed beds. The shoot appears after 4 months. About 50% of the seedlings are selected when they are 7 to 8 months old. The healthy seedlings are tranplanted in pits which are usually 3' × 3' × 3'. The pits are dug up 25 to 30 ft. apart about 1 to 3 months earlier and are manured and disinfected before the transplanting. The plant requires constant manuring. Wood ash (obtained by buring leaves, husks etc.) cattle manure and occasionally green manure are usually given during the rainy season. To obtain high yield potash supply is very much required. The nuts are harvested 6 to 8 times in a year.

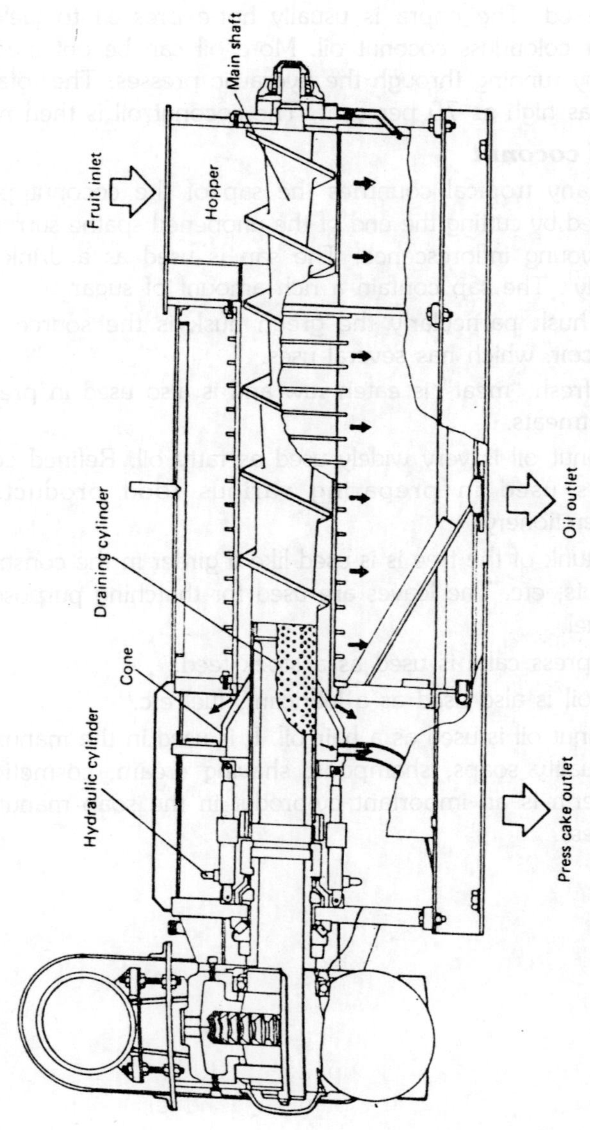

Fig. 17.8. A screw press, or expeller used to express oil from palm fruits.

Extraction of oil

The harvested nuts are cleared off the husks. The shell is split open by hand or by machine. The dried 'meat' is called *copra*. The copra or sometimes fresh 'meat' is crushed before oil is expressed. The copra is usually hot-expressed to yield pale yellow or colourless coconut oil. More oil can be obtained from residue by running through the hydraulic presses. The total yield of oil is as high as 70 per cent. The coconut oil is then refined.

Uses of coconut

1. In many tropical countries the sap of the coconut palm is tapped by cutting the end of the unopened spathe surrounding the young inflorescence. The sap is used as a drink called 'toddy'. The sap contain a rich amount of sugar.
2. The husk particularly the green husk is the source of the fibrecoir. which has several uses.
3. The fresh "meat" is eaten raw and is also used in preparing sweetmeats.
4. Coconut oil is very widely used as fatty oil. Refined coconut oil is used in preparing various food products and conferctionery.
5. The tunk of the tree is is used like a girder in the construction of huts, etc. The leaves are used for thatching purposes and as fuel.
6. The press cake is used as a stock feed.
7. The oil is also used as a lubricant, fuel etc.
8. Coconut oil is used as a hair oil. It is used in the manufacture of quality soaps, shampoos, shaving cream, cosmetic, etc. Glycerin is an important by-produt in the soap manufacture process.

18

ESSENTIAL OILS

'Essential oils' are a group of highly aromatic volatile organic substances, possed by several thousands species of plant of about 60 families. These differ from the fixed fatty oils of the plant in several ways. Unlike the fixed plant oils they do not give any oily feel and evaporate when they come in contact with air. Chemically they are terpenes which act as carriers of the aromatic substances. Terpenes are hydrocarbon in nature, being made up of isoprene units (C_5H_8), and existing as unsaturated straight chain molecules, or as ring structures, readily combining with other organic groupings.

Most essential oils also contain camphors, and the more odoriferous compounds present in them consist of oxygen derivatives of terpenes, alcohols, esters, aldehydes and ketones. Besides terpenes, essential oils may also be oxygenated and sulphuretted oils. The essential oil have smaller molecules, usually less than 20 carbon atoms long. The 'essential oils' are given the name during the Middle Ages because the pleasant smelling, highly volatile liquids were considered to be essential constituents of the plants.

Nature

The process of essential oil formation in the plants is still uncertain. They are believed to be by-products of metabolism and are found to be greatest in regions of photosynthetic activity. They are secreted in the plant body, or into internal glands which may develop in any part of the plant tissue, flowers (rose), fruits (orange), leaves (mint), bark cinnamon), rhizome (ginger), wood or seeds (cardamom), etc. Secreted oil is deposited in the intercellular spaces where an oil sac is gradually develpoed due to the breaking up of the cell walls. Chemically the oils may be different in different parts of the plant.

Function

The exact function of the essential oil is not known. They are believed to perform a number of functions; attract insects for pollination; repel animals and parasites : act as wound fluids; minimise the effect of heat on transpiration; act as hydrogen donors in oxidoreduction reactions; act as potential sources of energy and act as an antiseptic substance.

Uses

Essential oils are used for number of purposes. The most important use of the oils is in the manufacture of perfumes. They are used in perfuming soaps, deodorants and toilet preparations. Various food and beverage products and tobacco are flavoured using these oils. Essential oils are also used as stimulants and antiseptics, as constitutents of medicines, as a laboratory reagent, as solvents in the paint industry, as insecticides, and as a component of pastes, polishes, ink, glue, etc.

Extraction of Essential Oils

There are four methods for extracting the essential oils from the soild portion of the plant. There are distillation, expression, cold fat extraction and solvent expression.

Distillation

This is the most widely used method to extract the essential oils from various plant materials. However, it is not employed with flowers having delicate smell to prevent any destruction or chemical changes to oil which can be brought about by heat. Distillation in the case of immiscible oil is brought about with boiling water or steam. Boiling water or preferably steam is allowed to pass over the crushed material kept in a still when the essential oil is freed from the oil glands. The oil is vapourized along with water and the two are cooled and condensed in an adjoining condenser.

The oil or *essence* separates from water forming either an upper or lower layer depending upon its density. The oil is removed and filtered. This oil may be a mixture of essential oils. They can be separated or purified further by fractionation technique described below. Distillation in the case of miscible oil is by fractionation or rectification techniques. Plant materials or mixtures of essential oils obtained from steam distillation are distilled by this method. In this process the temperature is gradually increased when the more volatile oils distill first and the less volatile ones distill progressively later. The distillate can be further purified by distilling it again and again.

Expression

The rinds of citrus fruits are subjected to great pressure to squeeze out the essential citrus oils. Centrifugation is performed to separate the oil mixed with juices of the fruits. Hand operated presses or crushers or large mechanical presses are used to express the essential oils.

Fat Extraction

Cold fat extraction or *enfleurage* is done at normal temperature. This method is generally employed in the case of flowers with delicate oils which are adversely affected by distillation, etc. Pure tallow or lard which are odourless and are highly absorptive of floral essences are used to cover glass plates arranged one above the other in wooden frames.

The perfumed flower parts are placed on these fat layers for few days, after which they are replaced by a fresh batch of flowers. After several dozen successive batches of flowers (throughout the blooming season) are allowed to saturate the fat with the essence, the fat layer ('*pomade*') is subjected to alcoholic extraction. The essential oils being soluble in alcohol dissolve in it which is then concentrated to yield the perfume oil. Sometime *maceration* is employed, which is the process of fat extraction performed at high temperature. In this case small pieces of certain flowers are digested with hot oil or melted fat. The resulting pomades are subjected to alcoholic extraction.

Solvent Extraction

This is comparatively recent process employed in the last one hundred years only (since 1879). In this process the perfumes are extracted directly by means of certain purified volatile solvents like petroleum ether, benzol or alochol. The solvent is then allowed to evaporate away in vacuum leaving a semi-solid residue (*concrete*) of oils and waxes. The alcohol is added to remove wax as oils dissolve in it and the wax is filtered off. The wax is then filtered off. The alcohol is also then removed and a concentrated form of perfume oil is obtained.

Classification of the Essential Oils

A water-tight classification of essential oil plant is not possible since many of them yields oils which have uses in perfumery, in flavouring, in medicine as well as in industries. However, the essential oils are classified into three general groups for the purpose of convenience : (1) perfume oils, (2) flavouring (spice) oils, and (3) industrial, medicinal or special purpose oils.

Perfumes

Perfumes have a very old history and find references in the records of the earliest civilizations. The Egyptians and the Hebrews are known to have used them for personal and religious puroses. perfumes were used them for personal and religious purposes. Perfumes were used in the burial rites for Egyptian pharaohs. The Chinese used the perfumes on their body and essences for flavouring their food. Fragrant incense was used by the Chinese, Indian, Persian, Jews and the Babylonians as offerings to their gods. Perfumes used to be a part of religious functions of the Romans and Greeks. The rich Greeks used different perfumes for different parts of the body. At Versailles, King Louis XIV of France personally supervised the proper blending of esences for his royal bath, a particular essence formula for each day of the year.

Later due to a very widespread demand for essence in France, perfume industry was created by royal character. Today also France is the leading country in world in perfume industry followed by England, India, Turkey and United States. The natural perfumes are usually adulterated with synthetic ones to make them slightly cheaper. The synthetic materials are, however, not very lasting. The costly perfumes are usually blends of several natural perfumes.

The perfumes are normally mixed with less volatile oils, the fixatives, which may be of plant or animal origin (musk, ambergris, civet, etc.). To prevent the deterioration of perfumes due to oxidation and polymerization on coming in contact with air, the perfumes are stored in closed and completely filled containers.

Examples of Perfume Oils

1. *Lemon-grass oil*. It is obtained from the leaves of *Cymbopogon citratus* of the grass family, *Gramineae*. It is grown in India,

Sri Lanka, Congo. Madagascar, Guatemala, Haiti, Brazil, Salvador. etc.

2. *Saffron oil*. It is obtained from the dried stigmas of *Crocus sativus* of the *Iridaceae*. It is grown in the Mediterranean area and the Asiatic tropics.
3. *Jasmine*. The essence is obtained form the flowers of *Jasminum officinarum* var. *grandiflorum*. The plant is grown in southern France and other Mediterranean countries.
4. *Khus-khus oil*. (Vetiver oil). It is obtained from the roots of a grass. *Vetiveria zizanioides*. It occurs wild in many parts of India. It is cultivated in Java. Reunion. the West Indies and in the southern United States of America.
5. *Champaca oil*. It is obtained from the flowers of *Michelia champaca*. It is a famous perfume of India.
6. *Rose oil* (Otto, or Ottar, or Attar of roses). It is obtaind from flowers of damask rose, *Rosa damascena*, grown chiefly in Bulgaria. They are also grown in France, Italy, North Africa, Asia Minor and India.
7. *Sandalwood oil*. *Santalum album* and allied species are its sources.
8. *Geranium oil*. It is obtained from the leaves of several species of *Pelargonium* of the family *Geraniaceae*. They are chiefly grown in southern Europe and northern Africa.
9. *Narcissus oil*. It is obtained from *Narcissus jonquilla*, *N. poeticus* of *Amaryllidacae*. It is cultivated in southern Frnace.
10. *Neroli oil* or *Orange blossom oil*. Flowers of certain species of orange, particularly bitter orange (*citrus aurantium*) of *Rutaceae* are the main source of the oil.
11. *Cassie oil*. It is obtained from the flowers of *Acacia farnesiana* of the *Leguminosae*. It is chiefly grown in France, Algeria, Egypt, Syria and India.
12. *Lavender*. It is manufactured from the flowers of *Lavandula officinalis* of the *Labiatae*, which is grown in France and England.
13. *Yiang-yiang oil* (cananga oil). It is obtained from the petals of the flowers of *Cananga odorata* of the family *Annonaceae*. It is grown in the Philippines, Vietnam, Cambodia and other parts of the southeastern Asia.

Medicinal and Industrial Oils

Peppermint Oil

It is obtained from some species of *Mentha* (*M. paiperata*) of the *Labiatae*. It is indigenous to Europe. It is now cultivated in European countries, U.S.A., Japan and India and many other countries. The oil has a content of methanol. Chewing gums, sweets, tooth pastes, etc., are flavoured with it. It is of great medicinal value and is used as an antiseptic lotions and in digestive disorder.

Cedarwood Oil

It is obtained from the wood of eastern red Cedar (*Juniperus virginiana*) of the *Cupressaceae*. The plant is native to eastern United States. It is a cheap oil.It is used as a cleaning agent in preparing permanent slides. It is used with oil immersion lenses, in perfumery, soaps, dedorants and as an adulterant of expensive essential oils.

Ajowan Oil

It is manufactured from seeds of *Trachyspermum copticum* of the *Umbelliferae*. The plant is a small annual herb of India. It also grows in Egypt and Persia. The oil has a high thymol content. It is of medicinal value and is used in digestive disorders. The fruits are used as constitutents of curry powders.

Camphor and Camphor Oil

Camphor oil is liquid at higher temperatures, but on cooling the substance called camphor separates out. Camphor is a colourless, tough crystalline solid at oridinary temperature with a sharp odour and pungent, aromatic taste. It volatises very slowly. Celluloid and various nitrocellulose compounds are obtained from it and it is also of medicinal value. It is a part of religious functions in India. It is burnt before the deities of gods. Stems, twigs and leaves of the camphor tree, *Cinnamomum camphora* of the *Lauraceae* are the main sources of Comphor. The plants is native of China, Formosa and Japan.

The cultivation of the tree is now restricted to Formosa and Japan due to prodution of synthetic camphor in U.S.A. and England. The tree can grow well at elevations of up to 3,000 to 4,000 feet and between 44°N and 45°S latitudes. The plant is a large tree which can attain a height of a hundred feet or more. It is, however, of smaller height when cultivated. The camphor is deposited as a solid in the cells. Camphor is extracted from the plants which are 50 or more years old. The wood, twigs and leaves are macerated and distilled with steam. Sublimation with quick lime and charcoal is the method employed for further purifiction of crude crystalline camphor which seperates out from comphor oil on cooling. The residual oil contains an artificial sassafras oil, *safrole*, which is a more valuable product than camphor. It is used for the production of heliotropin.

INDEX

Q

R